49
Battery-Powered
Two-IC Projects
Delton T. Horn

TAB ## TAB BOOKS Inc.
Blue Ridge Summit, PA

FIRST EDITION
FIRST PRINTING

Copyright © 1989 by **TAB BOOKS Inc.**
Printed in the United States of America

Library of Congress Cataloging-in-Publication Data

Horn, Delton T.
    49 battery-powered two-IC projects / by Delton T. Horn.
       p.   cm.
    Includes index.
    ISBN 0-8306-9165-0   ISBN 0-8306-3165-8 (pbk.)
    1. Integrated circuits—Amateurs' manuals.   I. Title II. Title:
Forty-nine battery-powered two-IC projects.
TK9966.H66   1989
621.381'5—dc20   89-34639
                                                    CIP

TAB BOOKS Inc. offers software for sale. For information and a catalog,
please contact TAB Software Department, Blue Ridge Summit, PA
17294-0850.

Questions regarding the content of this book should be addressed to:

    Reader Inquiry Branch
    TAB BOOKS Inc.
    Blue Ridge Summit, PA 17294-0214

Acquisitions Editor: Roland S. Phelps
Book Editor: Steven L. Burwen
Production: Katherine Brown

Contents

Introduction

Electronics is becoming increasingly complex and sophisticated, even for hobbyists. At one time, believe it or not, an AM radio or a light dimmer was considered an advanced project. Today hobbyists are building robots, computers, and high-grade stereo and video equipment.

There is still a place for simpler projects, however, and that's where this book comes in. It's nice to be able to turn out a complete, working project in just a single evening. Beginners should learn their skills on relatively simple and inexpensive projects before tackling more complex projects. Even an advanced hobbyist can enjoy the relaxation of whipping together a quick-and-easy project. Besides, many of these projects are fun or useful in themselves.

Each project in this book is built around just two readily available integrated circuits (ICs). None of the projects should cost you more than about $15. Construction time for any of the projects should be well under two or three hours.

I hope you enjoy these simple, but interesting, two-IC electronics projects.

1

Introduction to the Projects

All of the projects in this book are relatively simple. Any one of them can easily be constructed in just an evening or two either by an experienced hobbyist or a beginner. Each project is built around just two inexpensive integrated circuits and a handful of external components.

This introductory chapter probably will be a review for many readers, especially those who have read the companion volume to this book, *49 Battery-Powered IC Projects* (TAB BOOKS No. 3155). This chapter offers a number of handy tips and shortcuts for building the projects in this book, or virtually any other electronic project you might come across.

FINDING THE COMPONENTS

You should have no problem finding any of the components for the projects in this book. In designing and choosing these projects, I have limited myself as much as possible to readily available parts.

Most of the parts can be purchased at a Radio Shack store that can be found just about everywhere. Most cities also have one or more independent parts supply houses.

Other good sources for electronics parts are the mail-order companies that advertise in the back of electronics hobbyist magazines. You can find such magazines at almost any newsstand. Of the mail-order companies, surplus houses frequently have the best bargains, if they happen to have what you need.

You must take the shipping and handling charges into consideration when determining which mail-order company offers the best buys. You can't just go by their price lists, which are often misleading. If company A charges an average of 2 to 10 cents more per component than company B, but has a lower shipping and handling charge, you might get more for your money from company B.

Sometimes you can find some great bargains by ordering grab bags of components. A grab bag is a "pig in the poke" assortment of parts. Be careful though. Grab bags are always sold "as is," so the reliability and honesty of the seller is a prime consideration in grab bags. Parts in grab bags usually are untested, so expect to get some duds.

Generally the best (and most reliable) bargains are to be had with classified grab bags. Rather than a general grab bag of who knows what, you get a grab bag of 100 resistors, or 50 capacitors, or something similar. Sure, you'll probably get some pretty oddball values, but you'll almost certainly get some items you can use. It isn't always possible, but try to stick with grab bags that are guaranteed to include marked components. Figuring out the values of a hundred or so unmarked components is usually more trouble than it's worth.

Grab bags are a good way to build up your "junk box." Every electronics hobbyist should have a junk box—a collection of common components that are likely to be useful in future projects. This cuts down the cost of your hobby considerably. Experienced hobbyists with well-stocked junk boxes probably will be able to build many of the projects in this book without spending a cent.

Another way to build up your junk box is by acquiring and dismantling discarded equipment. Even if it doesn't work at all, it is still bound to have a number of perfectly good parts. When you "cannibalize" old equipment, concentrate on the more expensive components like switches, semiconductors, and potentiometers. Don't bother with less expensive parts like resistors and disc capacitors. Removing them from the circuit is more trouble than it's worth, and the leads are likely to be too short to be reusable. If the equipment is very old, leave the electrolytic capacitors. The dielectric could be dried out. Incidentally, be very careful around any large capacitors. Even if no power has been applied to the circuit for quite awhile, they might still hold a considerable charge. If you're not careful, you could receive a painful, and possibly dangerous, electrical shock.

You can also replenish your junk box by dismantling your old projects that you're no longer using. If you made a mistake on a project and it didn't work, it probably still has some perfectly good, reusable components. Don't throw them away.

Keep your junk box well organized. If you just throw everything into a shoebox, it probably won't do you much good. You'll never be able to find the part you need when you need it. (Somehow it always manages to turn up in your junk box right after you've ordered a new one.)

Get a multidrawer parts cabinet, or make some kind of homemade dividers. Separate your junk-box components by type. If you have a large number of a given type of component (such as resistors and capacitors), you'll also want to subdivide them by value. You don't need a separate container for every individual value. Organize these parts by a value range.

Your storage compartments don't have to be anything elaborate or expensive. I have stored small parts in ordinary business envelopes. This proved handy, because the type of part enclosed and the value (or value range) could be marked directly on the envelope.

MAKING SUBSTITUTIONS

Occasionally you won't have exactly the right part on hand for a project. In some projects, certain components are very critical, but usually substitutions can be made. The projects in this book are fairly flexible, and you shouldn't have any problems making substitutions. In many cases, you are encouraged to experiment with other component values. If a value does happen to be critical, this will be mentioned in the text for that project.

Always make sure that any substitute component has voltage, current, and/or power ratings equal to or greater than those of the part being replaced. If a ¼-watt resistor is called for (as in most of the projects in this book), you can certainly use a ½-watt resistor. It will be slightly larger physically, but that probably won't be important in most hobbyist applications. On the other hand, you should *not* replace a ½-watt resistor with a ¼-watt unit. It might not be able to stand up to the power fed through it by the circuit. If a resistor burns out or changes value because of overheating, your project won't work properly.

Similarly, if a 100 μF, 15-volt electrolytic capacitor is called for, feel free to use a 100 μF, 25-volt device. If the original diode has a PIV rating of 100 volts, you'll have no problem using a diode with a PIV rating of 200, or even 300, volts.

These power ratings are the minimum values for the component used in the circuit. As a rule, you should use a component close to the called-for ratings. Components with higher ratings tend to be physically larger, and—more important to the hobbyist—more expensive.

There is one important exception to this type of overrating of replacement components: fuses. Never replace any fuse with one that has a higher current rating. The current rating for a fuse is not selected arbitrarily. The fuse is used to protect the other components in the circuit. It is rated for the maximum amount of current that can be drawn safely by the circuit. If you use a larger value fuse, an expensive transistor or IC could blow out to protect the fuse, which pretty much defeats the whole purpose.

To substitute semiconductors (transistors and ICs) you will need a suitable substitution guide. These are available from various publishers and component manufacturers. It is best to have a number of substitution guides from a variety of sources. Many manufacturers market a line of components specifically designed as general substitutes. For example, there is the Motorola HEP series, and the Sylvania ECG line.

A general-purpose substitute will work in most of the projects in this book. If the exact component is required, this will be noted in the text. To the best of my knowledge, there are no substitutes available for some of the ICs used in these projects.

When using a substitute component, always be sure to double-check pin numbering for ICs or the lead arrangement for transistors. Sometimes a suitable electrical substitute won't be an exact mechanical substitute. In some cases you might have to modify the circuit wiring slightly.

Resistor Substitutions

Most substitutions will be made with passive components, such as resistors and capacitors. If you don't have the exact value called for, you might have something in your junk box that is close enough to do the job.

Most passive component values are pretty flexible within a specific range or tolerance. For example, resistors are available with the following tolerance ratings: 20%, 10%, 5%, 1%, and 0.1%. The actual value of the component might be off from the stated value up to the tolerance percentage of the stated value. As an example, let's assume we have a 100k (100,000 ohm) resistor at each of the tolerances just listed. The maximum error and the range of values for each tolerance are shown in Table 1-1. Note that even a 20 percent tolerance resistor might be exactly at its stated value. The tolerance rating simply specifies the maximum error range. The component is guaranteed to be somewhere within the range defined by the tolerance rating.

Generally, 20 percent tolerance resistors are used only in the most non-critical applications. They are becoming increasingly rare.

Very tight tolerance resistors, such as 1 percent or 0.1 percent units, tend to be relatively expensive, so they are used only in those applications where very high precision is necessary.

For most applications, 10 percent or 5 percent tolerance resistors are your best bet. This rule works for all of the projects in this book. Unless noted otherwise, you can use either 5 percent or 10 percent tolerance resistors in these projects.

You can always substitute a component with a tighter tolerance rating than the one called for in the parts list. For example, if a 10 percent unit is called for, you can certainly use a 5 percent device.

You can go in the other direction too, if you use an ohmmeter to determine the actual value of the resistor. You can use a 20 percent resistor in place of a 10 percent unit, if the measured value is no more than 10 percent off from the nominal, stated value.

If you don't have the exact stated value handy, you can often substitute a resistor with a value that is close. For example, if the

Table 1-1. Resistor Tolerances.

Rating	Maximum Error	Minimum Actual Value	Maximum Actual Value
20%	±20000	80,000	120,000
10%	±10000	90,000	110,000
5%	±5000	95,000	105,000
1%	±1000	99,000	101,000
0.1%	±100	99,900	100,100

project calls for a 4.7k, 10 percent resistor, you can probably get away with using a 3.9k resistor, especially if it had a 5 percent tolerance. To be on the safe side, you should breadboard the circuit with such a substitution to make sure it works properly before permanently soldering the components in place.

You can also make up unavailable values (including unusual values) by combining resistors in series and in parallel. You don't have to use a single component to give the desired resistance.

For resistors in series (Fig. 1-1), the resistances are simply added together:

$$R_t = R1 + R2 + \ldots R_n$$

For parallel combinations, like the one illustrated in Fig. 1-2, the reciprocal of the total is equal to the sum of the reciprocals of each of the component resistances. This sounds a lot harder than it really is. It is much clearer in equation form:

$$1/R_t = 1/R1 + 1/R2 + \ldots 1/R_n$$

For series combinations, the total effective resistance is always greater than any of the component resistances. On the other hand, for parallel combinations, the total effective resistance is always smaller than any of the component resistances.

You can use a modified form of the parallel equation when there are just two resistances in parallel:

$$R_t = (R1 \times R2)/(R1 + R2)$$

Fig. 1-1. Resistances in series add.

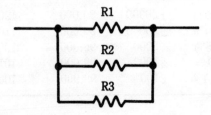

Fig. 1-2. Resistors also can be combined in parallel.

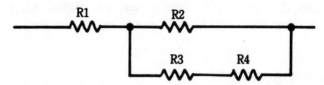

Fig. 1-3. Series and parallel resistances can be combined in a single circuit.

Series and parallel resistances also can be combined as shown in Fig. 1-3.

Capacitor Substitutions

Capacitors can be substituted in a manner similar to resistors. Capacitor tolerances (especially for large electrolytic capacitors) tend to be much wider than resistance tolerances, so you have much more room for making substitutions. If the stated value is reasonably close, it probably will work just fine, unless a high-precision capacitor specifically is called for.

Capacitor markings vary somewhat, so components that are identical essentially can be marked differently by different manufacturers. For example, there is rarely any noticeable difference between disc capacitors marked 0.2 μF or 0.22 μF, or even 0.25 μF. In many applications, you could even substitute a capacitor marked 0.15 μF.

Capacitors can be combined in series or parallel, just like resistors. The rules for determining the total effective value, however, are the exact opposite of those used with resistors. The formula for capacitors in series is:

$$1/C_t = 1/C1 + 1/C2 + \ldots 1/C_x$$

For parallel capacitances, the formula is:

$$C_t = C1 + C2 + \ldots C_x$$

Because of the wider tolerances of capacitors, the equations are much less precise for capacitors. If you can get the total effective capacitance reasonably close to the desired value, it probably will be perfectly acceptable.

Series and parallel capacitances can be combined in a single circuit (Fig. 1-3). You can combine them in series to produce values greater than, or combine them in parallel to produce values less than, either component above to obtain the value you want.

BREADBOARDING

In the early days of electronics, experimenters mounted components on a piece of wood or a breadboard. The term *breadboarding* has survived to describe any form of solderless, temporary construction of an electronic circuit.

Today breadboarding generally is done on a special solderless socket, like the one shown in Fig. 1-4. The socket is covered with rows of holes that component leads can fit into. The holes are spaced to accommodate the pins of the ICs in DIP-type housings.

Within the socket, the holes are connected internally in a specific pattern. Figure 1-5 shows a common interconnection pattern.

Fully integrated breadboarding systems also are available. In addition to the solderless socket, the system includes commonly used circuit modules, such as variable power supplies and oscillators. You won't have to build these auxiliary circuits each time they are needed. Conveniently mounted switches and potentiometers also are provided.

It is usually a good idea to breadboard any circuit before constructing it permanently. You might decide the circuit doesn't really do what you want, so you will want to be able to reuse the components in some other project. Problems also are easier to correct in a breadboarded circuit. Suppose you happen to have a dud IC. It is easy to replace it in a solderless socket, but it would be a nuisance to change it if you had to desolder it and solder in a new one.

Occasionally an error can occur in a schematic diagram, despite the best efforts of editors and technical writers. If you breadboard the circuit first, you will find out that there is no problem, and it will be a lot easier to find and correct the erroneous or missing connection.

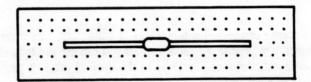

Fig. 1-4. Modern breadboarding is done on a special solderless socket.

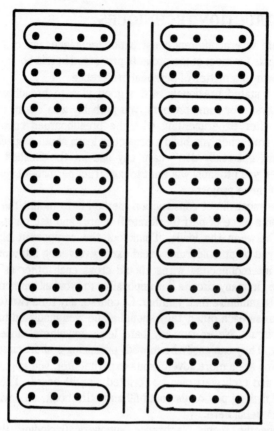

Fig. 1-5. A typical interconnection pattern for a solderless socket.

Best of all, breadboarding permits you to experiment and easily make modifications in the circuit. You can make sure any non-identical substitution works. You might want to see how circuit performance changes if a certain resistance is increased, or if a certain capacitance is decreased. Perhaps you might want to add another output device. You'll need to find out if this addition loads down the circuit excessively. If you breadboard the project first, you can find out the answers to all such questions before you have to commit yourself with the soldering iron.

CONSTRUCTION TECHNIQUES

There are several different construction techniques you can use. All are suitable for most of the projects presented in this book. It really boils down to a matter of personal preference.

As I stated previously, it's a good idea to breadboard each project first. For some projects, you probably won't want to go any further after experimenting with the breadboarded circuit. This often is true when you build a project to learn about a specific type of circuit, but really don't have any use for the finished project. However, you probably will want to construct permanent versions of many of these projects.

The most direct construction method is to mount the components on a piece of perf (perforated) board, and use point-to-point wiring between the leads. Think out the component placement before you begin soldering. Avoid excessively long interconnecting wires. Interconnecting wires should cross each other as little as possible. In many circuits it is impossible to avoid all wire crossings, but try to minimize their number. Of course, any time two (or more) wires cross each other, they should all be well insulated to avoid a short circuit. Never run any bare leads anywhere close to one another. They can easily be moved, producing a short circuit.

If you are a more adventurous experimenter, you might want to design and etch your own pc (printed circuit) board. The ins and outs of making pc boards could fill an entire book, so I will not go into it in detail here.

Recently, many electronics suppliers (including the ever-present Radio Shack) have started offering universal pc boards. There are pc boards with a generalized copper pattern already etched onto them, somewhat similar to the interconnection pattern of a solderless socket. It often takes a little more work to arrange the components to get the connections right on a universal pc board, but this method of circuit construction is a nice compromise between a customized pc board and point-to-point perf-board construction.

PRECAUTIONS

In any type of circuit construction, you should consider using sockets for the ICs. A socket is a relatively cheap form of insurance. An IC is a semiconductor, of course, and is therefore heat sensitive. Soldering all of those closely spaced pins could easily cause the chip

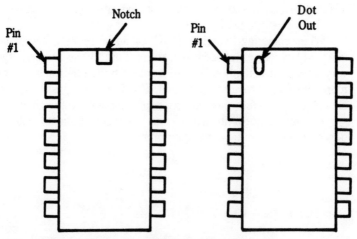

Fig. 1-6. Most ICs have a marking to help you locate pin #1.

to over-heat and become damaged. If you do solder an IC directly into a circuit, the use of a soldering heat sink is practically mandatory.

Some people in the field feel IC sockets are usually wasteful. In many cases, a socket can cost more than the IC being protected. This is true, but if you do damage an IC in direct soldering (or get a bad one, which happens occasionally), you will have to desolder it, then solder in a replacement. This is a lot of tedious, extra work. A socket can save you quite a bit of frustration.

Always remember never to make <u>any</u> changes in a circuit with the power connected. This especially includes putting an IC into a socket or taking it out. <u>Always</u> remove power first.

In some very-high-frequency circuits, a socket can interfere with proper operation of the circuit (especially if a cheap socket is used). However, this book does not include any such "fussy" circuits. Unless you do a lot of work in the upper radio bands or in high-speed digital systems, you probably will never run into such problems.

Whether or not you are using sockets, be very careful with the orientation of all ICs. Don't install an IC backwards. Also, you must make sure none of the pins get bent up under the body of the IC, where they won't make electrical contact with the circuit. Make it a habit to always double-check such potential problems before applying power to any project. To help in their orientation, most ICs have a clear marking on the front end, or an indication of the location of pin #1, as illustrated in Fig. 1-6.

When soldering (whether an IC, or an IC socket), always be aware of how closely spaced adjacent pins are to one another. It is very easy inadvertantly to create a solder bridge between two or more pins if you are not very careful. This is another possibility you should always double-check before applying power to the circuit.

Any specialized precautions for individual projects will be described in the text, where appropriate.

2

Switching Circuits

Switching is one of the most basic electronic functions. This chapter features projects that perform various automated switching and related tasks.

PROJECT 1: FOUR-STEP SEQUENCER

The circuit shown in Fig. 2-1 is a four-step event sequencer. Four devices are switched on and off automatically in sequence.

The circuit uses five timers, in two ICs. Four of the timers are contained in a 558 quad timer IC. The fifth timer is a separate 555 IC. The parts list for this project is given in Table 2-1.

In operation, one, and only one, of the four outputs is activated at any time. First output A is turned on. When it is turned off, output B is turned on. Output C is turned on after output B is timed out. When output C goes off, then output D is turned on. After output D goes off, output A is activated again, and the whole cycle is repeated. The on time for each output is individually adjustable via potentiometers R3, R8, R9, and R10. Potentiometers R1 and R11 interact to set the overall sequence rate.

Almost any circuit can be controlled with this circuit. By using the outputs to drive suitable relays or SCR circuits, virtually any electrically powered device can be driven by this project. The potential applications are limited only by your imagination.

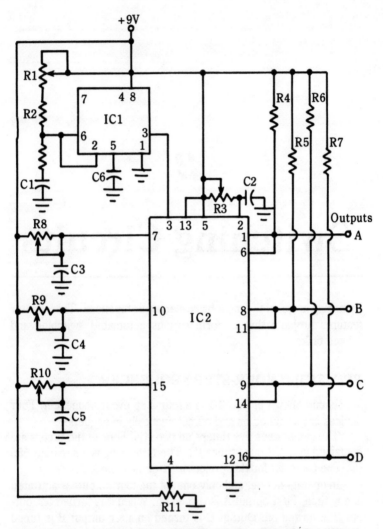

Fig. 2-1. The sequencer circuit in Project 1 steps through four steps.

PROJECT 2: DELAYED TRIGGER TIMER

Timer, or monostable multivibrator, circuits are useful in many control applications. The timer's output is normally low. When a trigger pulse is received, the output immediately goes high and remains in that state for a fixed period of time. (In some circuits, the low and high states are reversed.)

Table 2-1. Parts List for Project 1.

Component	Part
IC1	555 timer
IC2	558 quad timer
C1	0.0047-μF capacitor
C2 – C5	0.1-μF capacitor
C6	0.01-μF capacitor
R1	500k potentiometer
R2	2.2k, ¼-watt resistor
R3, R8 – R11	10k potentiometer (or trimpot)
R4 – R7	3.3k, ¼-watt resistor

In some applications, you might not want the timing cycle to begin as soon as the trigger pulse is received. Instead, you might want the timing cycle to begin at some specific time after the trigger signal. In other words, the trigger is effectively delayed.

The circuit for the delayed trigger timer project is shown in Fig. 2-2. Note that this circuit is built around two 555 timers in the

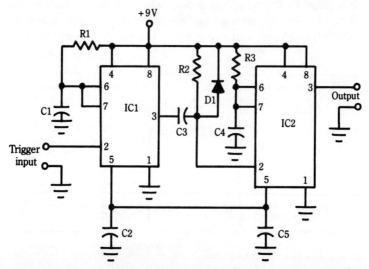

Fig. 2-2. This dual timer circuit in Project 2 features a delayed trigger.

monostable mode. IC1, and its associated components, controls the delay time (T1) between the actual input pulse and the beginning of the output pulse. IC2, and its associated components, controls the length of the actual output pulse (T2). The action of this circuit is illustrated in Fig. 2-3.

The delay time for this circuit is set by the values of R1 and C1. The normal 555 monostable equation is used:

$$T1 = 1.1 \times R1C1$$

Similarly, the output pulse time is controlled by the values of R3 and C4.

A typical parts list for this project is given in Table 2-2. The timing components are marked with asterisks. You can experiment with other values for these components.

If you use the component values in the parts list, the delay period is 5.17 seconds, and the output pulse time is 0.451 second. In other words, the output goes high 5.17 seconds after the trigger pulse is received. The output reverts to its normal low state 5.621 seconds after the trigger pulse is received. You can try other component values to create other timing periods.

PROJECT 3: DUTY-CYCLE CONTROLLER

Rectangle waves, which are widely used in switching applications, have two possible states—high and low. Theoretically, the signal switches instantly from one state to the other. The actual wave-

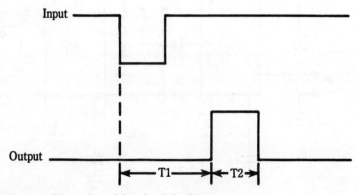

Fig. 2-3. The action of the circuit in Fig. 2-2.

16

Table 2-2. Parts List for Project 2.

Component	Part
IC1, IC2	555 timer
D1	1N914 diode
C1	10-μF, 25-volt electrolytic capacitor*
C2, C5	0.01-μF capacitor
C3	0.001-μF capacitor
C4	0.5-μF capacitor*
R1	470k, ¼-watt resistor*
R2	10k, ¼-watt resistor
R3	820k, ¼-watt resistor*

*Timing component—experiment—see text

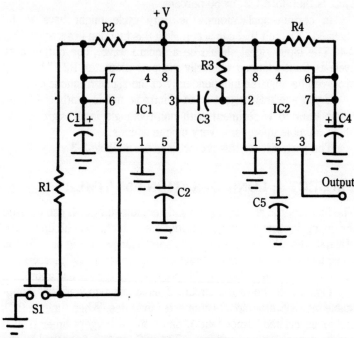

Fig. 2-4. This circuit controls the duty cycle of a rectangle wave in Project 3.

Table 2-3. Parts List for Project 3.

Component	Part
IC1	555 timer
IC2	TL507 A/D converter
C1	5000-pF capacitor
C2	0.01-μF capacitor
R1, R3	4.7k, ¼-watt resistor
R2	100k potentiometer

forms have some finite transition time between states, but it is usually negligible, and can be ignored in practical applications.

The portion of each cycle that is in the high state is called the *duty cycle*. A square wave is high exactly half of each cycle. The duty cycle is therefore 1:2, or 50 percent.

In certain applications, the duty cycle might have to be changed. A circuit that can accomplish this function is shown in Fig. 2-4. The input signal should be a square wave. The output duty cycle is then adjusted manually via potentiometer R2. The duty cycle can be adjusted independently of the signal frequency.

If you use this project in an electronic music synthesizer, you might want to experiment with non-rectangle input signals. You might stumble upon some very unique sounds.

The parts list for this project is given in Table 2-3.

PROJECT 4: LONG-DURATION 555 TIMER

The 555 timer is an easy to use and versatile device. It can be used for timing applications ranging from a fraction of a second up to several minutes. However, what if you need a longer timing period? You could use a more expensive timer circuit, or you could cascade two (or more) 555 timer stages. This project uses the second approach.

Figure 2-5 shows the circuit. This is just two standard monostable multivibrator circuits connected in series. When timer A (IC1 and its associated components), times out, it triggers timer B (IC2 and its associated components). The 555 timer is triggered by the falling edge of a pulse, so timer B does not begin its cycle until after timer A has timed out.

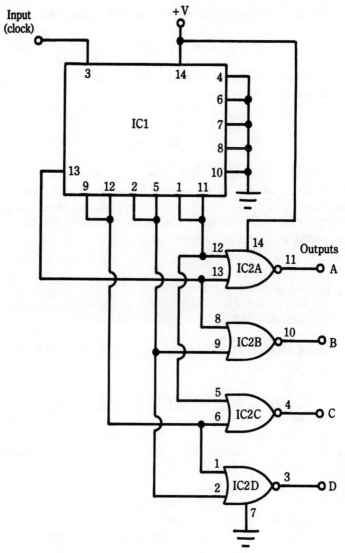

Fig. 2-5. Two timer stages can be cascaded to obtain a longer timing period in Project 4.

Timer A's time period is set by the values of resistor R2 and capacitor C1:

$$T_A = 1.1 \times R2C1$$

Similarly, R4 and C4 set the time period for timer B:

$$T_B = 1.1 \times R4C4$$

The total time period of the circuit as a whole is simply the sum of the two individual stages:

$$\begin{aligned} T_t &= T_A + T_B \\ &= 1.1 \times R2C1 + 1.1 \times R4C4 \end{aligned}$$

A typical parts list for this project appears in Table 2-4. The components marked with asterisks (*) are timing components. I encourage you to experiment with other values for these components. Using the component values given in the parts list, the timing period works out to a nominal value of:

$T_t = 1.1 \times R2C1 + 1.1 \times R4C4$
   $= (1.1 \times 470000 \times 0.00025) + (1.1 \times 2200000 \times 0.0001)$
   $= 129.15 + 242$
   $= 371.25$ seconds
   $= 6$ minutes, 11.25 seconds

Table 2-4. Parts List for Project 4.

Component	Part
IC1, IC2	555 timer
C1	250-μF, 25-volt electrolytic capacitor*
C2, C3, C5	0.01-μF capacitor
C4	100-μF, 25-volt electrolytic capacitor*
R1, R3	10k, 1/4-watt resistor
R2	470k, 1/4-watt resistor*
R4	2.2-megohm, 1/4-watt resistor*

*Timing component—see text

This circuit can be set up for reliable time periods of up to 20 or 30 minutes. It's just a matter of selecting the right timing components.

PROJECT 5: CLOCKED SEQUENCER

This project is somewhat similar to Project 1, which was also a four-stage sequencer. The big difference here is that an external (rather than internal) clock signal is used. This offers greater versatility in certain applications when the sequencer must be operated in step with other circuitry.

The schematic for this project is shown in Fig. 2-6. The simple parts list appears in Table 2-5. Note that just two ICs are called for. No discrete components are used in this project.

This sequencer has four outputs. One, and only one, of the outputs will be high at any given time. The other three outputs will

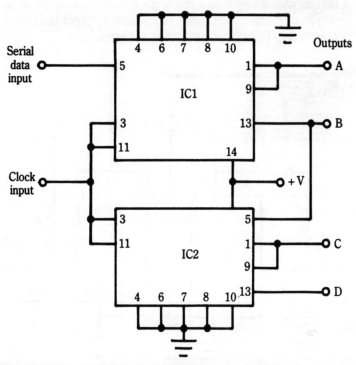

Fig. 2-6. This sequencer circuit in Project 5 is controlled by an external clock.

Table 2-5. Parts List for Project 5.

Component	Part
IC1	CD4013 dual D-type flip-flop
IC2	CD4001 quad NOR gate

be high at any given time. The other three outputs will all be low. Each time an input (clock) pulse is detected by this circuit, the current output is switched off (goes low), and the next output in sequence is activated (goes high).

This project can be used in any application in which multiple events must occur in sequence.

PROJECT 6: SHIFT REGISTER

A shift register is very similar to a sequencer. It is actually a specialized type of counter. Shift registers are used for short-term memory, or to step through a programmed sequence.

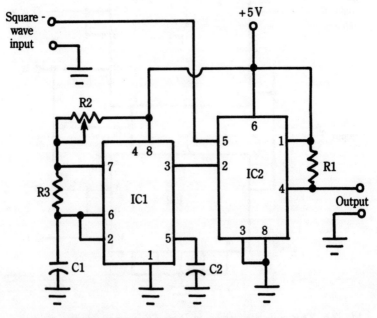

Fig. 2-7. A shift register is a form of sequencer.

In this circuit, illustrated in Fig. 2-7, any bit pattern (1's and 0's) can be fed serially into the data output, that is, one after another. One input bit is accepted on each controlling clock pulse. The data is then fed out in parallel fashion four bits at a time at the outputs (A through D). On each clock pulse, everything is moved over one place to make room for the new input bit.

As you can see, this circuit has four outputs and two inputs. One of the inputs is the data bit input, and the other is the clock-pulse input.

The operation of this shift register can best be understood with an example. Suppose the following bit pattern was being fed into the circuit:

$$1 - 0 - 0 - 1 - 1 - 0 - 1 - 1 - 1 - 0 - 0 - 0 - 0-$$

and just 0s from then on. Assuming that all of the outputs started out low (all 0s), the following output pattern would occur:

Output			
A	B	C	D
1	0	0	0
0	1	0	0
0	0	1	0
1	0	0	1
1	1	0	0
0	1	1	0
1	0	1	1
1	1	0	1
1	1	1	0
0	1	1	1
0	0	1	1
0	0	0	1
0	0	0	0
0	0	0	0

The outputs will remain all 0s until more data is input.

The data can be taken off serially (instead of in parallel) by using just one of the outputs. If output D is used, the data is delayed by a

Table 2-6. Parts List for Project 6.

Component	Part
IC1, IC2	CD4013 dual D-type flip-flop

number of clock pulses equal to the number of stages in the shift register. There are four stages in this circuit.

Table 2-6 shows the parts list. This is probably the simplest parts list in this book. Only two identical ICs are needed to construct this project. No external components are required.

3

Amplifiers

The most basic and frequently encountered of all electronic circuits is the amplifier. Virtually all electronic systems include at least one amplifier stage. An amplifier accepts an input signal and produces an output with a different (usually larger) level, or amplitude.

It shouldn't be at all surprising that the projects featured in this chapter are amplifiers of various types.

PROJECT 7: POWER AMPLIFIER

Many amplifier devices are now available in IC form. Although there have been vast improvements in the power-handling capabilities of semiconductors, there are still practical limitations. The more power an amplifier has, the more heat it has to disperse. All semiconductors, including ICs, are heat-sensitive. Too much heat can damage or destroy the delicate semiconductor crystal.

There are some tricks for upping the output power without causing the ICs to self-destruct. One such technique is to use two amplifier ICs in tandem, as illustrated in Fig. 3-1. A parts list for this project appears in Table 3-1.

This circuit is built around a pair of LM380 audio amplifier chips. By using two amplifiers in this bridge configuration, you can double the voltage gain across the load for a given supply voltage. In effect, this increases the power capability by a factor of four over a single LM380.

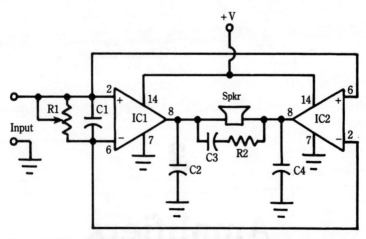

Fig. 3-1. The two LM380 amplifiers in Project 7 can be used together to achieve a higher output power.

Table 3-1. Parts List for Project 7.

Component	Part
IC1, IC2	LM380 audio amplifier
C1	50-pF capacitor
C2 – C4	0.1-μF capacitor
R1	2-megohm potentiometer
R2	2.7-ohm, ½-watt resistor
Spkr	8-ohm speaker

For a power-amplifier application like this project, external heat-sinking of the ICs is virtually mandatory.

PROJECT 8: PREAMPLIFIER

The circuit shown in Fig. 3-2 is a preamplifier circuit that can be used in many audio-frequency applications. Table 3-2 shows the parts list. It can be used in radios, musical-instrument amplifiers, or

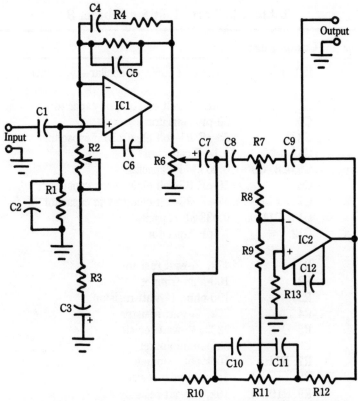

Fig. 3-2. This preamplifier circuit in Project 8 can be used in many audio applications.

stereo systems (a complete preamplifier circuit is needed for each channel in the system), among other applications. The brute-force power amplification is performed by a separate power-amplifier stage.

Circuit gain is set by potentiometer R6. Between the two op-amp stages is a passive filter network that allows the user some manual control over the frequency response of the circuit. The high-frequency (treble) range can be cut back somewhat by adjusting potentiometer R7. Similarly. potentiometer R11 controls the low-frequency (bass) response.

Trimpot R2 is adjusted for minimum distortion in the output signal. An oscilloscope would be a big help in making this adjustment. Alternatively, you could use a sine wave as the test signal, and adjust R2 for the purest sounding tone.

Table 3-2. Parts List for Project 8.

Component	Part
IC1, IC2	op amp (TL080, or similar—see text)
C1	1-μF, 35-volt electrolytic capacitor
C2	50-pF capacitor
C3	68-μF, 35-volt electrolytic capacitor
C4	0.1-μF capacitor
C5, C8, C9	0.0033-μF capacitor
C6	10-μF, 35-volt electrolytic capacitor
C7	47-μF, 35-volt electrolytic capacitor
C10, C11	0.033-μF capacitor
C12	15-pF capacitor
R1	47k, ¼-watt resistor
R2	100-ohm trimpot
R3	100-ohm, ¼-watt resistor
R4	27k, ¼-watt resistor
R5	220k, ¼-watt resistor
R6	5k potentiometer
R7, R11	100k potentiometer
R8	3.3k, ¼-watt resistor
R9, R10, R12	10k, ¼-watt resistor
R13	68k, ¼-watt resistor

For best results, high-quality, low-noise, op-amp ICs such as the TL080 should be used. If the application is not critical, almost any op amp, even the lowly 741, can be used, but it might introduce too much hiss and noise into the output signal for any true, high-fidelity applications. In a preamplifier, the signals are at a very low level. Any noise is amplified and can quickly reach an unacceptable level.

PROJECT 9: DIGITALLY PROGRAMMABLE AMPLIFIER

This project, which is illustrated in Fig. 3-3, is rather unique. It is an analog (inverting) amplifier circuit with programmable gain. That is, the gain of the amplifier can be controlled by a four-bit digital value applied to control inputs A through D.

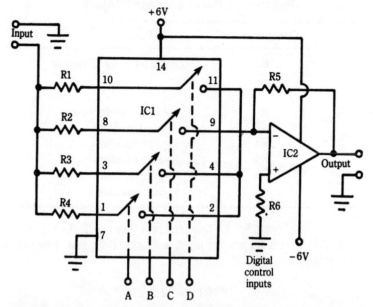

Fig. 3-3. The gain of this analog amplifier in Project 9 is digitally programmable.

The secret of this circuit is a special CMOS digital IC called the CD4066 quad bilateral switch. This device is sometimes known as a *quad analog switch* because analog signals can be fed through the internal switches. Only the control inputs must be digital in nature.

A digital input for each internal switch controls its position. A logic 0 (low) opens the appropriate switch, while a logic 1 (high) closes it.

In the circuit of Fig. 3-3, the switches select the input resistor(s) that the analog input signal (V_i) passes through. If more than one switch is closed at one time, the multiple resistors are in parallel. The selected value of the input resistance combines with the value of feedback resistor R5 to determine the gain of the inverting amplifier:

$$V_o = V_i \times (R5/R_i)$$

where V_o is the output signal, V_i is the input signal, and R_i is the selected input resistance value.

Table 3-3. Parts List for Project 9.

Components	Part
IC1	CD4066 quad bilateral switch
IC2	op amp
R1	2.2k, ¼-watt resistor
R2	4.7k, ¼-watt resistor
R3	10k, ¼-watt resistor
R4	18k, ¼-watt resistor
R5	33k, ¼-watt resistor
R6	15k, ¼-watt resistor

A typical parts list for this project is given in Table 3-3. Resistors R1 through R4 are the input resistors. Note that their values are staggered:

$$R1 = 2.2\text{ k}$$
$$R2 = 4.7\text{k}$$
$$R3 = 10\text{k}$$
$$R4 = 18\text{k}$$

Each input resistor is approximately twice the value of its preceding one.

There are 16 possible control input combinations, from 0000 to 1111. If the digital control inputs are all 0s (0000), then all four switches are open, preventing any input signal from reaching the op amp. The circuit effectively is cut off. There could be some minor leakage in some cases. An open CD4066 switch is not a true open, but a very high resistance.

Control input D is the least-significant bit. Control input A is the most-significant bit. Table 3-4 shows the gain for each possible digital control input combination. Some of the gain values are rounded off. The negative signs simply indicate that the output signal is inverted (polarity reversed) from the input signal.

One disadvantage of this circuit is the gain does not increase with the digital control input value. This could be somewhat awkward in some applications. Still, this circuit can prove to be extremely useful for controlling amplifier gain by computer.

Of course, you are encouraged to experiment with other resistor values in this project.

Table 3-4. Gains for Control Inputs.

Control	R_i	Gain
00000	∞	off
0001	2200	−15
0010	4700	−7
0011	1499	−22
0100	10000	−3.3
0101	1803	−18.3
0110	3197	−10.3
0111	1303	−25.3
1000	18000	−1.8
1001	1960	−16.8
1010	3727	−8.9
1011	1383	−23.9
1100	6429	−5.1
1101	1639	−20.1
1110	2715	−12.2
1111	1215	−27.2

PROJECT 10: MULTIPLIER AMPLIFIER

In a sense, any amplifier is a multiplier circuit. A variable (the input voltage at any given instant) is multiplied by a constant (the amplifier's gain). For instance, if the gain is 10, the output voltage (V_o) is always equal to the input voltage (V_i) multiplied by ten. That is:

$$V_o = V_i \times 10$$

Suppose you want to multiply two variable values, you want to use two input voltages (V_i1 and V_i2), and you want the output (V_o) to be the product of the two inputs. That is:

$$V_o = V_i1 \times V_i2$$

You could use a voltage-controlled amplifier (VCA) (see Project 12). A totally different but equally effective approach involves the use of logarithmic amplifiers.

If you know how to use an old-fashioned slide rule, you are already familiar with logarithms, whether you realize it or not.

Today mechanical slide rules have been replaced almost totally by pocket calculators that are more efficient and versatile.

Basically, a logarithm of a number is the exponent that indicates the power to which a specific base number must be raised to produce the given number. Any value can be used as the base. In the type of logarithm known as the *common logarithm*, the base is ten. For our purposes here we will be using only common logarithms. A base of ten is therefore always assumed.

Logarithms can be a little confusing and intimidating when you first encounter them. The definition seems awfully complex. Fortunately, things become a little clearer with a few examples.

The common logarithm of 100 is 2:

This is the same as saying that the base (10) raised by a power of 2 equals 100:

$$10^2 = 100$$

This relationship can be expressed in general terms as follows:

$$\log x = y$$
$$10^y = x$$

Because we are using only common logarithms, the base is assumed to be 10.

Not all logarithms are nice, neat whole numbers like 2. Most include digits to the right of the decimal point. Here are a few more common logarithms:

$$\log 1 = 0$$

(The logarithm of 1 is always 0, regardless of the base.)

$$\log 2 = 0.3010$$
$$\log 3 = 0.4771$$
$$\log 4 = 0.60206$$
$$\log 5 = 0.6987$$
$$\log 10 = 1.0000$$
$$\log 12 = 1.0792$$
$$\log 20 = 1.3010$$

$$\log 25 \quad = 1.3979$$
$$\log 30 \quad = 1.4771$$
$$\log 1000 \quad = 3.0000$$
$$\log 10,000 \ = 4.0000$$
$$\log 100,000 = 5$$

Raising ten to the logarithm reproduces the original number. Note that the logarithm is always significantly less than the original value.

The logarithm is not defined for zero or negative numbers. Negative logarithms correspond to values between zero and one. For example:

$$\log 0.1 \quad = -1.0000$$
$$\log 0.2 \quad = -1.6990$$
$$\log 0.005 = -3.3010$$

Each logarithm consists of two parts, called the *characteristic* and the *mantissa*. These two portions of the logarithm are divided by the decimal point. The characteristic appears to the left of the decimal point, and the mantissa to the right. Published log tables typically give only the mantissas, because these values will repeat for different characteristics, according to simple rules. The characteristic indicates how many times the number can be divided by a whole number power of the base.

This is clearer when we look at some specific examples:

$$\log 2 \quad = 0.3010$$
$$\log 20 \quad = 1.3010$$
$$\log 200 \quad = 2.3010$$
$$\log 2000 \ = 3.3010$$

Note that the mantissa is the same in each of these examples. Two cannot be divided by 10 (the base), so the characteristic is 0. In the next example, however, 20 equals 2 (1×10), so there is a characteristic of 1. Similarly, 200 is 2 $\times (10 \times 10)$, and 2000 is 2 $\times$ (ten $\times$ ten $\times$ ten), so the characteristics are 2 and 3, respectively.

The opposite of the logarithm (log) is the antilogarithm (antilog). An antilogarithm takes a logarithmic value and returns the original number. For example:

$$\log 2 = 0.3010$$
$$\text{antilog } 0.3010 = 2$$

At this point, you probably are wondering what the point of this mathematics lesson is. At first glance, logarithms seem like a lot of unnecessary fuss and bother. Why should we waste our time with them?

The reason is that logarithms can simplify a lot of relatively complex mathematical functions. For example, consider the process of multiplication. Some multiplication problems are simple enough. For example, few of us would have any problem multiplying 2 times 3, or 11 times 22. However, what if you had to work any of the following problems?

$$53,789 \times 683,090 = ?$$
$$37,000,000,000 \times 54,100,000 = ?$$
$$0.0000005502 \times 0.000000991004 = ?$$
$$4,170,540,000 \times 0.000000392 = ?$$

You could work these problems directly with a pencil and paper, but it would be a lot of work, and it would be very easy to make a mistake.

Logarithms can make problems like these a lot easier. If you add the logarithms of two numbers and then take the antilogarithm of the sum, you will end up with the product of the two original numbers. In algebraic form this is:

$$\text{antilog (log A + log B)} = A \times B$$

The use of logarithms permits multiplication via simple addition. This is also useful electronically because a circuit that multiplies directly would be difficult to design, but logarithmic (and antilogarithmic) amplifiers and summing (addition) amplifiers are easy enough to come up with.

To prove that the logarithm method of multiplication really works, let's use a trivial example. Of course you know that 2 multiplied by 3 equals 6. Let's see if logarithms will give us the same result:

$$
\begin{aligned}
C &= A \times B \\
C &= \text{antilog (log A + log B)} \\
&= \text{antilog (log 2 + log 3)} \\
&= \text{antilog (0.30103 + 0.47712)} \\
&= \text{antilog 0.77815} \\
&= 6
\end{aligned}
$$

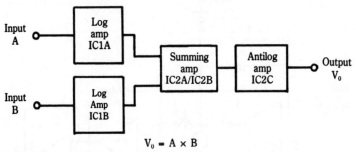

$$V_0 = A \times B$$

Fig. 3-4. Logarithms can be used to perform multiplication.

Yes, we get the right answer to the problem. This method will work for any multiplication problem.

A block diagram for an electronic multiplier circuit is shown in Fig. 3-4. You should be able to see how each stage in the circuit corresponds directly to a part of the logarithmic multiplication formula:

$$\text{antilog } (\log A + \log B)$$

The actual circuit appears in Fig. 3-5. Five op-amp stages are used. To keep the IC count down to two chips, we can use a dual op amp and a quad op amp. The parts list for this project is given in Table 3-5.

Table 3-5. Parts List for Project 10.

Component	Part
IC1	dual op amp (747, or similar)
IC2	quad op amp (324, or similar)
	(one stage not used)
Q1, Q2, Q3	npn transistor (2N2222, or similar)
D1, D2, D3	diode (1N4001, or similar)
C1, C2, C3	.01-μF capacitor
R1, R5, R13	47k, ¼-watt resistor
R2, R6	330-ohm, ¼-watt resistor
R3, R7	10k trimpot
R4, R8	2.2k, ¼-watt resistor
R9, R10, R11, R12	10k, ¼-watt resistor

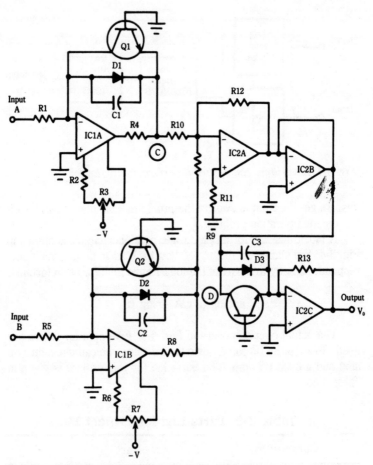

Fig. 3-5. Project 10 is a multiplying amplifier circuit.

Don't let the apparent complexity of this circuit throw you. Remember, it is made up of the four simple stages illustrated in Fig. 3-4. Each op amp and its associated components make up one stage, with one exception. The summing amplifier stage is made up of two op amps. This is because IC2A is an inverting summing amplifier. IC2B is an inverting buffer to correct the polarity of the sum signal. These two op amps combine to form a non-inverting summing amplifier. Except for these considerations, everything in this circuit is perfectly straightforward.

The input signals to the circuit should be kept relatively small. Remember, if the product of the two input voltages exceeds the circuit's supply voltage, the output will be clipped. Multiplication can increase values very quickly.

Trimpots R3 and R7 are used to calibrate the logarithmic amplifier stages.

For maximum clarity, power-supply connections to the op-amp ICs are omitted in the schematic diagram. Don't forget to make these connections in building the project. The power-supply connections always are assumed.

PROJECT 11: EXPONENTIAL AMPLIFIER

Another application for logarithms is in calculating exponents.

Sometimes it is necessary to raise an input signal to a specific power. This process amounts to multiplying the number by itself a specified number of times. Consider 5 raised to the fourth power as an example:

$$5^4 = 5 \times 5 \times 5 \times 5$$
$$= 625$$

This is a simple enough example, but sometimes using exponents is not so easy. Consider the following examples:

$$1753^{43} = ?$$
$$8.09^{89} = ?$$
$$7057.223^{9.17} = ?$$

I certainly wouldn't want to figure manually any of these problems.

Logarithms can come to the rescue and simplify things enormously. To raise any number A to the nth power, you can use this logarithmic formula:

$$A^n = \text{antilog}(n \times \log A)$$

Note that n is used directly in this formula. You do not take the logarithm of n.

You can check out this formula by solving for 5 to the fourth power again:

$$5^4 = \text{antilog } (4 \times \log 5)$$
$$= \text{antilog } (4 \times 0.69897)$$
$$= \text{antilog } 2.79588$$
$$= 625$$

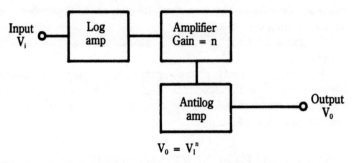

$$V_0 = V_i^n$$

Fig. 3-6. Logarithms also can be used to calculate exponents.

This technique also can be applied electronically using op amps. A block diagram for a simple exponential amplifier is shown in Fig. 3-6. Because the exponential power (n) is often a constant in many practical applications, the circuitry can be simplified, and you can use the gain of single non-inverting amplifier to perform the multiplication. The gain of this stage is equal to n. If your application requires a single variable exponent value, you can combine this project with Project 10. Of course, this hybrid project will require more than two ICs.

The complete schematic for this project appears in Fig. 3-7. Table 3-6 shows the parts list. For most practical applications, high-grade op-amp ICs are not required.

Table 3-6. Parts List for Project 11.

Component	Part
IC1, IC2	dual op amp (747, or similar)
Q1, Q2	npn transistor (2N2222, or similar)
D1, D2	diode (1N4001, or similar)
C1, C2	0.1-μF capacitor
R1, R6	47k, ¼-watt resistor
R2, R7	330-ohm, ¼-watt resistor
R3, R8	10k trimpot
R4	2.2k, ¼-watt resistor
R5	25k potentiometer (adjust gain-n)

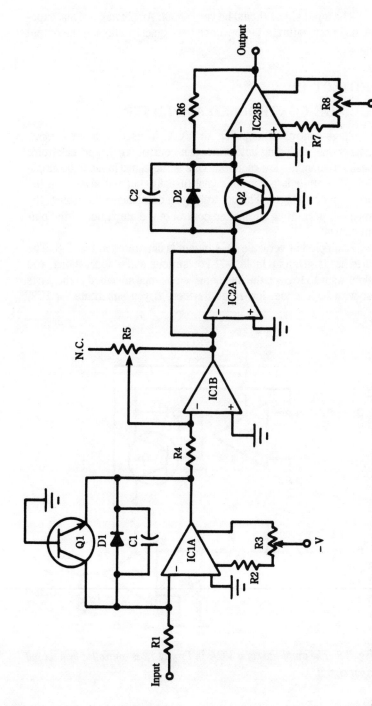

Fig. 3-7. Project 11 is an exponential amplifier circuit.

The input signal should be very small. Amplifying a signal exponentially can saturate the op amps very quickly, clipping the output signal.

PROJECT 12:
VOLTAGE-CONTROLLED AMPLIFIER

A voltage-controlled amplifier, or VCA, is used in a great many applications, including automatic gain control (agc), and electronic music. Two inputs are required. One is the signal input to be amplified. The other is a control signal (usually, but not always, in the form of a dc voltage). The control signal determines the gain of the amplifier, permitting automated control of the amplitude of the output signal.

A simple but versatile VCA circuit is illustrated in Fig. 3-8. The parts list is given in Table 3-7. For serious audio applications, you might want to replace the inexpensive op amps specified in the parts list with high-grade, low-noise devices. If you substitute for IC2,

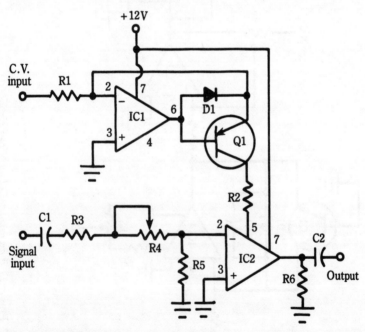

Fig. 3-8. The amplitude of a VCA in Project 12 is controlled by a second input signal.

Table 3-7. Parts List for Project 12.

Component	Part
IC1	op amp (741, or similar)
IC2	op amp (CA 3080, or similar)
Q1	pnp transistor (2N3905, or similar)
D1	diode (1N914, or similar)
C1, C2	0.1-μF capacitor
R1, R2, R3, R5, R6	10k, ¼-watt resistor
R4	100k potentiometer

the replacement op amp should have an equivalent extra input like the one at pin #5 of the CA3080.

IC1 is a voltage-to-current converter, while IC2 is a current-controlled amplifier. Combining the two stages produces a fine voltage-controlled amplifier.

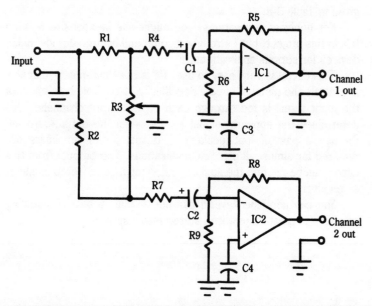

Fig. 3-9. Project 13 permits you to pan any mono signal between two stereo outputs.

41

Table 3-8. Parts List for Project 13.

Component	Part
IC1, IC2	op amps (see text)
C1, C2	1-μF, 35-volt electrolytic capacitor
C3, C4	0.1-μF capacitor
R1, R2, R4, R7	10k, ¼-watt resistor
R3	10k potentiometer
R5, R8	33k, ¼-watt resistor
R6, R9	18k, ¼-watt resistor

PROJECT 13: PAN-POT AMPLIFIER

In stereo systems, a signal can be placed anywhere between the two speakers. Moving the apparent signal source from speaker to speaker is known as *panning*.

This project, shown in Fig. 3-9, allows you to pan any mono signal between two stereo outputs. The parts list for this project is given in Table 3-8.

For hobbyist applications, you might use inexpensive op-amp ICs in this project, but I strongly recommend high-grade, low-noise devices for serious applications.

This circuit is very easy to use. R3 is the "pan-pot"—that is—it controls the panning effect. Specifically, it controls how much of the input signal is fed through each of the output channels. This determines the apparent signal source when the outputs are fed through a pair of loudspeakers. External power amplifiers are required for almost all practical applications. The outputs from this circuit can be fed into the auxiliary (aux) inputs of a stereo amplifier or receiver.

You can combine several pan-pot amplifiers with a summing amplifier to create a powerful stereo mixer system.

4

Oscillators and Signal Generators

The projects in this chapter all generate ac signals of various types and waveshapes, ranging from the simple to the complex. Oscillators, along with amplifiers and power supplies, are among the most commonly encountered types of electronic circuits. These projects are useful in a variety of applications.

PROJECT 14: DELUXE
RECTANGLE-WAVE GENERATOR

Many circuits generate rectangle waves, which often are used to control the duty cycle, or pulse width. Unfortunately, such control often interacts with other parameters, particularly frequency. When the duty cycle is altered, the signal frequency also is changed. This is obviously undesirable in many applications.

The circuit shown in Fig. 4-1 offers independently adjustable frequency and duty cycle. Two separate 555 timer stages are used. One controls the duty cycle, while the other controls the signal frequency. The parts list for this project appears in Table 4-1.

Potentiometer R1 is used to control the frequency. The duty cycle is set by potentiometer R4. The duty cycle can be changed without affecting the signal frequency.

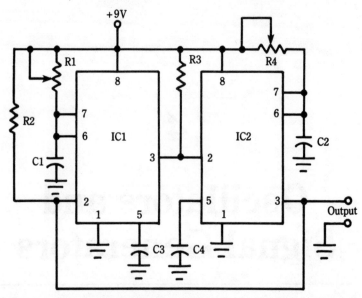

Fig. 4-1. This rectangle-wave generator of Project 14 features independently adjustable frequency and duty cycle.

Table 4-1. Parts List for Project 14.

Component	Part
IC1, IC2	555 timer
C1	0.1-μF capacitor
C2 – C4	0.01-μF capacitor
R1, R4	100k potentiometer
R2, R3	3.3k, ¼-watt resistor

PROJECT 15: STEPPED-WAVE GENERATOR

The circuit shown in Fig. 4-2 generates an unusual waveform known as a *stepped wave*. This waveform is made up of multiple rectangle waves. On an oscilloscope, the waveshape resembles a staircase. There are a number of possible variations of stepped

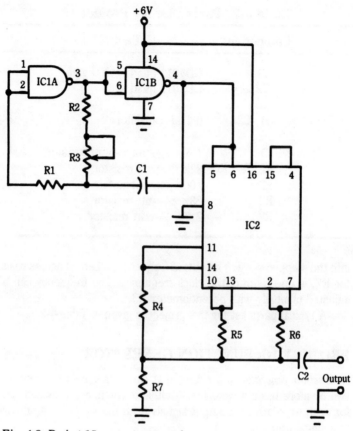

Fig. 4-2. Project 15 generates stepped waves.

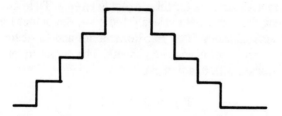

Fig. 4-3. This is the output waveform from the circuit of Fig. 4-2.

waves. This project generates a four-stage up/down stepped wave, as illustrated in Fig. 4-3.

IC1 and its associated components serves as a high-speed clock source. IC2 converts the clock signal (a simple square wave)

Table 4-2. Parts List for Project 15.

Component	Part
IC1	CD4011 quad NAND gate
IC2	CD4012 quad latch
C1, C2	0.1-μF capacitor
R1	1.5-megohm, ¼-watt resistor
R2, R4, R6	22k, ¼-watt resistor
R3	100k potentiometer
R5	33k, ¼-watt resistor
R7	2.2k, ¼-watt resistor

into the stepped wave. The output frequency, taken off across resistor R7, is one-tenth of the clock frequency. The frequency can be adjusted manually with potentiometer R3.

A typical parts list for this project is given in Table 4-2.

PROJECT 16: FUNCTION GENERATOR

A *function generator* is a circuit that can simultaneously generate two or more basic waveforms. Multiple outputs are provided, one for each waveform. All output signals from the various outputs are at a single frequency.

A simple two-output function generator project is shown in Fig. 4-4. A typical parts list for this project is given in Table 4-3.

Using the component values listed here, the output frequency will be approximately 1000 Hz. Resistor R4 and capacitor C2 are the frequency-determining components. The output frequency can be determined with this formula:

$$F = 5/(2 \text{ R4} \times \text{C2})$$

pi is a mathematical constant equal to approximately 3.14. The output frequency can be made manually adjustable by using a potentiometer for R4.

The two outputs available from this circuit are in the form of square waves (A) and triangle waves (B).

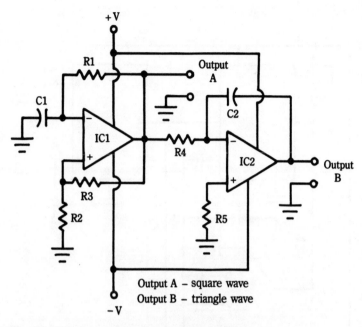

Fig. 4-4. *This function generator of Project 16 has two outputs.*

Table 4-3. Parts List for Project 16.

Component	Part
IC1, IC2	op amp (741, or similar)
C1, C2	0.1-μF capacitor
R1	47k, ¼-watt resistor
R2	100k, ¼-watt resistor
R3	10k, ¼-watt resistor
R4	1.5k, ¼-watt resistor
R5	4.7k, ¼-watt resistor

PROJECT 17: ODD-WAVEFORM GENERATOR

The circuit shown in Fig. 4-5 uses two 555 timers to generate a wide variety of oddball waveforms. Table 4-4 shows the parts list. These non-standard waveforms can be used in electronic music and similar applications.

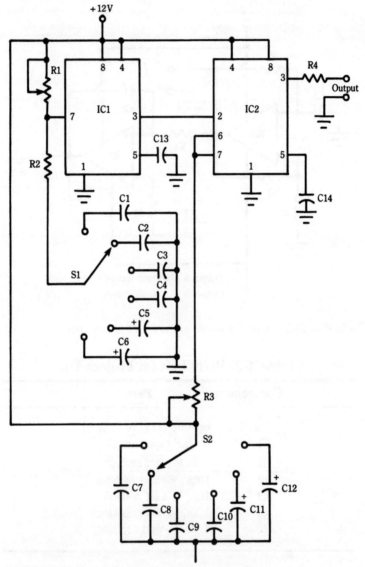

Fig. 4-5. Project 17 generates some very unusual waveforms.

By adjusting the two potentiometers (R1 and R3), and selecting various capacitors with switches S1 and S2, you can generate a wide range of unique sounds and effects. Some of the results will be very musical and pleasant; others will be totally dissonant. If nothing

Table 4-4. Parts List for Project 17.

Component	Part
IC1, IC2	555 timer
C1, C7, C13, C14	0.01-μF capacitor
C2, C8	0.047-μF capacitor
C3, C8	0.1-μF capacitor
C4, C9	0.47-μF capacitor
C5, C11	1-μF, 25-volt electrolytic capacitor
C6, C12	5-μF, 25-volt electrolytic capacitor
R1, R3	100k potentiometer
R2, R4	1k, ¼-watt resistor
S1, S2	single-pole, 6-position rotary switch

else, you should have a lot of fun playing with this circuit. Try experimenting with other component values. This project might not be the most useful project in this book, but it should be good for a few hours of fun.

PROJECT 18: BIRD CHIRPER

For some reason, the chirping of a bird has always been a popular electronic sound effect. Perhaps this is because it happens to be one of the easiest sounds to synthesize electronically. During Christmas time, you're likely to see ornaments that produce a warbling, birdlike sound. By using the circuit shown in Fig. 4-6, you can build your own chirping ornaments. The commercial units use similar circuitry.

None of the components are particularly critical in this circuit. Table 4-5 shows the parts list. You are quite free to experiment with various component values. If you change the component values, the results might not sound much like a chirping bird, but some very interesting sounds can be generated. In particular, I'd recommend trying different values for capacitors C1 and C2. These capacitors should be an electrolytic type within the 15 μF to 500 μF range.

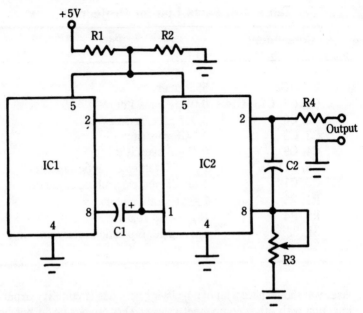

Fig. 4-6. Project 18 simulates the chirping of a bird.

Table 4-5. Parts List for Project 18.

Component	Part
IC1, IC2	LM3909 LED flasher/oscillator
C1	33-μF, 15-volt electrolytic capacitor
C2	0.1-μF capacitor
R1	3.3k, ¼-watt resistor
R2	1.5k, ¼-watt resistor
R3	25k potentiometer
R4	47-ohm, ¼-watt resistor

PROJECT 19: SIMPLE ORGAN-DRUM
SYNTHESIZER

This project is another fun one. The circuit, which is shown in Fig. 4-7, is a simple toy organ. The parts list appears in Table 4-6.

This project is built around the MK50240 top-octave generator IC (IC2). This chip accepts a clock signal at its input (from the oscil-

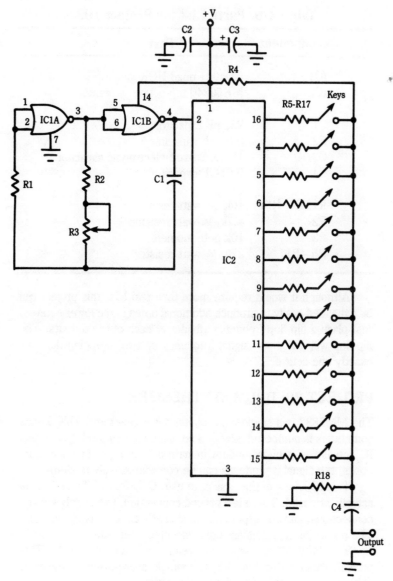

Fig. 4-7. Project 19 is a circuit for a simple toy organ.

lator circuit built around IC1), and produces multiple outputs corresponding to one full octave of an equally tempered musical scale. All of the notes will always be perfectly in tune with one another. Potentiometer R3 serves as a master tuning control.

Table 4-6. Parts List for Project 19.

Component	Part
IC1	CD4001 quad NOR gate
IC2	MK50240 top-octave generator
C1	250 pF capacitor
C2	0.01-μF capacitor
C3	10-μF, 25-volt electrolytic capacitor
C4	0.01-μF capacitor
R1, R4, R18	10k, ¼-watt resistor
R2	4.7k, ¼-watt resistor
R3	10k potentiometer
R5 – R17	27k, ¼-watt resistor

Although it would require more than two ICs, this project can be expanded easily to produce additional notes in the lower octaves. Just place a flip-flop frequency divider at each of the outputs. The flip-flop will divide the signal frequency by two, dropping the tone exactly one octave.

PROJECT 20: DRUM SYNTHESIZER

The MM5837 is a single-chip, digital noise generator. This device sometimes is numbered S2668. The noise pattern produced by this IC actually is pseudo-random in nature. For most practical purposes, the signal it produces can be considered truly random.

This IC is one of the easiest to use. Only four of the eight pins actually are used. There is a ground connection, two supply voltage connections, and an output terminal. That's all there is to this chip. The remaining four pins are not connected internally.

The MM5837 is useful for creating percussive sounds. This project, illustrated in Fig. 4-8, is a simple snare-drum synthesizer. Table 4-7 shows the parts list for this project.

Because the basic sound is built around noise, a high-grade op amp is not required. The lowly 741 will do just fine.

Briefly depressing switch S1 produces a sound that resembles a beat on a snare drum. If the switch is held down for more than a fraction of a second, you will get an unusual sound effect that does not sound drumlike.

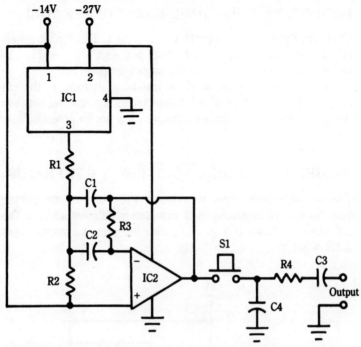

Fig. 4-8. The sound of a snare drum can be synthesized with this circuit for Project 20.

Table 4-7. Parts List for Project 20.

Component	Part
IC1	MM5837 noise generator
IC2	op amp (741, or similar)
C1, C2	0.02-μF capacitor
C3, C4	0.1-μF capacitor
R1	3.9k, ¼-watt resistor
R2	1k, ¼-watt resistor
R3	7.5k, ¼-watt resistor
R4	1.5k, ¼-watt resistor
S1	normally open, SPST pushbutton switch

PROJECT 21: FIRST TONE-BURST GENERATOR

There isn't too much to say about this project. This circuit generates bursts of tone, separated by brief periods of silence. The circuit is shown in Fig. 4-9. Table 4-8 shows the parts list.

Two controls are provided for manual adjustment of the circuit's output. Potentiometer R3 determines the spacing between the tone bursts, while the tone frequency is set by potentiometer R4.

PROJECT 22: SECOND TONE-BURST GENERATOR

This project is quite similar in function to the preceding one, but the tone bursts are produced by a completely different circuit. The schematic is shown in Fig. 4-10, with the parts list appearing in Table 4-9.

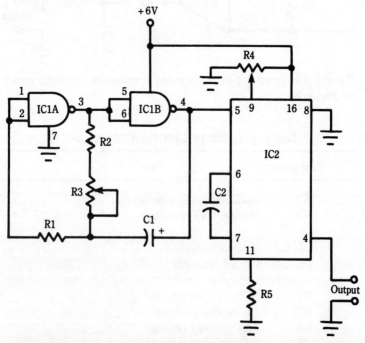

Fig. 4-9. This circuit for Project 21 produces bursts of tone separated by periods of silence.

Table 4-8. Parts List for Project 21.

Component	Part
IC1	CD4011 quad NAND gate
IC2	CD4046 digital phase-locked loop (PLL)
C1	3.3-μF, 35-volt electrolytic capacitor
C2	0.001-μF capacitor
R1	3.3-megohm, ¼-watt resistor
R2	47k, ¼-watt resistor
R3	250k potentiometer
R4	500k potentiometer
R5	100k, ¼-watt resistor

This tone-burst generator is built around two 555 timers. Timer A (IC1) turns the output tone on and off, while timer B (IC2) produces the tone itself.

Resistors R1 and R2 can be replaced by potentiometers to provide manual control over the output. Resistance R1 controls the burst rate, while R2 can be changed to adjust the tone frequency.

Normally timer A (IC1) operates at a sub-audible frequency, so separate bursts of tone will be heard. Some very unusual sounds

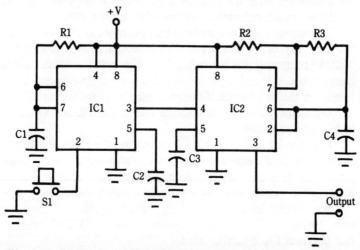

Fig. 4-10. Project 22 is a second tone-burst generator circuit.

Table 4-9. Parts List for Project 22.

Component	Part
IC1, IC2	555 timer
C1, C4	0.1-μF capacitor
C2, C3	0.01-μF capacitor
R1	330k, ¼-watt resistor
R2	12k, ¼-watt resistor
R3	4.7k, ¼-watt resistor
S1	normally open SPST pushbutton switch

can be generated by operating timer A at an audible frequency. Try reducing the value of capacitor C1. For a slower burst rate, try using a 1-μF or 10-μF electrolytic capacitor in place of C1.

PROJECT 23: COMPLEX TONE GENERATOR

Simple signal generators produce basic waveforms, such as triangle waves, rectangle waves, or ramp waves. Sometimes, however, more complex waveforms can be useful. This is especially true in electronic music, although there are other applications.

A complex signal generator circuit is shown in Fig. 4-11. The parts list appears in Table 4-10. This project is built around the XR2240 programmable timer IC. This chip includes a timer and a binary counter. It has eight outputs. Each successive output has a frequency that is one-half of the frequency of its predecessor.

Digital gates are used to combine four of the XR2240's outputs into a complex pattern.

Resistor R1 and capacitor C1 determine the signal frequency. Because multiple frequencies are being combined into a single complex signal, it is not easy to calculate the actual output frequency. Experiment with different values for these components.

You might also want to try creating different complex waveshapes by using different combinations of XR2240 outputs. Any pin from 1 to 8 may be used.

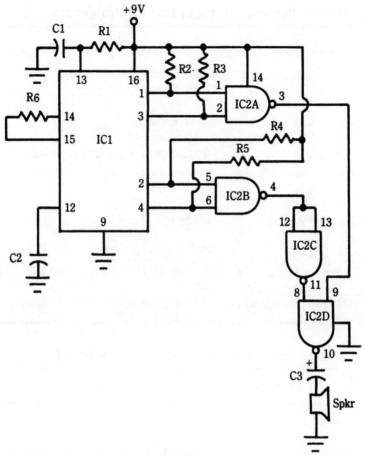

Fig. 4-11. Various complex waveforms can be generated by this circuit for Project 23.

PROJECT 24:

PROGRAMMABLE SIGNAL GENERATOR

Project 23 is capable of putting out a variety of unusual waveforms. Unfortunately, if you want to change the waveshape, you will have to physically rewire the circuit.

Figure 4-12 shows the circuit for this project. The parts list is given in Table 4-11.

This circuit is programmable. The waveshape can be changed dynamically just by adjusting one or more manual controls (potentiometers R1 through R9—odd numbers only).

Table 4-10. Parts List for Project 23.

Component	Part
IC1	XR2240 programmable timer
IC2	CD4011 quad NAND gate*
C1	0.1-μF capacitor**
C2	0.01-μF capacitor
C3	10-μF, 35-volt electrolytic capacitor
R1	47k, ¼-watt resistor**
R2, R3, R4, R5	10k, ¼-watt resistor
R6	22k, ¼-watt resistor
Spkr	small, 8-ohm speaker

For a different effect, substitute a CD4001 quad NOR gate. The
pin numbering is the same.
**Timing component—experiment with other values.

Table 4-11. Parts List for Project 24.

Component	Part
IC1	CD4017 decade counter
IC2	op amp (741, or similar)
C1, C2, C3, C5	0.01-μF capacitor
C4, C8	0.5-μF capacitor
C6	0.05-μF capacitor
C7	0.1-μF capacitor
R1, R3, R5, R7, R9, R11, R13, R15, R17, R19	1-megohm potentiometer
R2, R4, R6, R8, R10, R12, R14, R16, R18, R20	470k, ¼-watt resistor
R21, R22	22k, ¼-watt resistor
S1	5-position, single-throw rotary switch

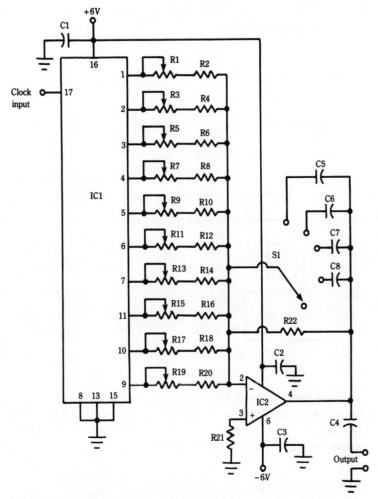

Fig. 4-12. Project 24 generates programmable stepped waves.

The circuit requires a digitally compatible (CMOS) input signal with a frequency ten times higher than the desired output frequency.

IC1 is a ten-step counter. The amplitude of each output is independently adjustable via the appropriate potentiometer. The result is a complex stepped wave, like the one illustrated in Fig. 4-13. This combination waveform is created by blending the ten counter outputs. Op-amp IC2 is a mixer for this purpose. If a capacitor is switched into the feedback network of the op amp, it will serve as a

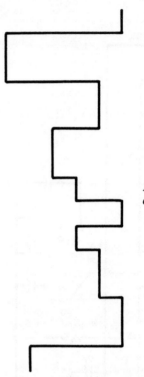

Fig. 4-13. *This is a typical waveform generated by the circuit of Fig. 4-12.*

low-pass filter, smoothing out the steps of the waveform. Switch S1 allows you to select between several capacitors (or no capacitor at all), offering various degrees of filtering. The capacitor value determines the cut-off frequency of the filter.

PROJECT 25: MUSICAL TONE GENERATOR

I'm not really sure about what this project should be called. It's a musical instrument, but not quite like most other musical instruments. In any case, it's fun to play. The schematic is shown in Fig. 4-14, and the parts list appears in Table 4-12.

The two sections of IC1 are the main signal generator, producing a rough sawtooth wave. The signal frequency can be adjusted with potentiometer R1. A calibrated knob would be helpful. A tone is fed to the output only when switch S1 is closed. R1 can be reset to a new note while the switch is open, that is, between notes.

IC2 is a modified VCA (voltage-controlled amplifier—see Chapter 3). Potentiometer R13 controls the amplitude envelope of the

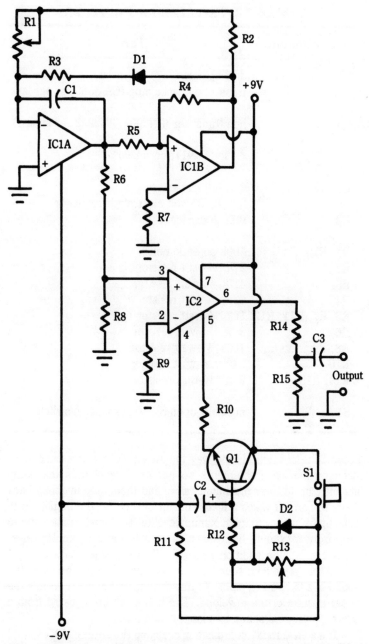

Fig. 4-14. This circuit is an unusual musical instrument.

Table 4-12. Parts List for Project 25.

Component	Part
IC1	dual op amp (747, or similar)
IC2	transconductance amp (CA 3080, or similar)
Q1	npn transistor (2N3904, 2N2222, or similar)
D1, D2	diode (1N914, or similar)
C1	0.12-μF capacitor
C2	3.3-μF, 35-volt electrolytic capacitor
C3	0.5-μF capacitor
R1, R13	100k potentiometer
R2	27k, ¼-watt resistor
R3	4.7k, ¼-watt resistor
R4	62k, ¼-watt resistor
R5, R7, R11, R12	10k, ¼-watt resistor
R6	3.9k, ¼-watt resistor
R8, R9	100k, ¼-watt resistor
R10	220k, ¼-watt resistor
R14, R15	2.7k, ¼-watt resistor
S1	normally open SPST pushbutton switch

notes. The envelope is the way the amplitude changes over time. Acoustic musical instruments do not produce sounds that jump instantly up to their maximum value, and then jump instantly back down to zero. It takes a finite amount of time for the amplitude to build up from zero to its maximum level (attack), and a finite time to drop from maximum back down to zero (decay). (Complex envelopes can include additional stages.) The envelope plays a major part in the way a note sounds. It is the main difference between the sound of a flute and a guitar. In this circuit, R13 mainly controls the decay portion of the envelope. The notes can be adjusted from a short, percussive sound to a long, legato tone.

It's a little tricky coordinating changing the setting of R1 (pitch) and the opening and closing switch S1, but with a little practice, you will be able to play simple tunes with this project.

5

Test Equipment and Measurement Circuits

Anyone who works in electronics, whether professionally or as a hobby, has to make numerous measurements. Standard test equipment, such as VOMs (volt-ohm-milliammeters) and oscilloscopes can be used to make almost any type of measurement. Some measurements are more convenient to make with these devices than others. Often, using an expensive generalized piece of test equipment for certain measurements is overkill. This chapter features specialized test equipment and measurement projects for a variety of purposes.

PROJECT 26: ABSOLUTE-VALUE METER

In some applications, the magnitude of a voltage is important, but not the polarity. For instance, I might need to detect five-volt signals of either polarity. The most obvious solution would be to use a pair of voltage detectors, one for +5 volts, and the other for −5 volts. This approach works, but it is not very elegant, and it often proves to be unnecessarily expensive.

A more practical solution is to use a circuit that can take the absolute value of the input voltage. Essentially, when we take the absolute value, the polarity sign is simply discarded and ignored. In algebra, the absolute value is indicated by two bars on either side of the number, that is:

$$|x| = \text{absolute value of x}$$

The absolute-value function always has a positive result, regardless of the polarity of the original number. Here are some typical examples:

$$|5| = |5|$$
$$|-7| = |7|$$
$$|4.32| = |4.32|$$
$$|-0.1| = |0.1|$$
$$|19| = |19|$$
$$|-19| = |19|$$
$$-|3| = |-3|$$

Did that last one catch you off guard? If the polarity sign is outside the absolute value bars, it is not affected by the function.

Figure 5-1 shows the circuit for the absolute-value indicator project. The parts list appears in Table 5-1.

This circuit is built around a pair of op amps. For most practical applications, inexpensive chips like the 741 can be used. A high-

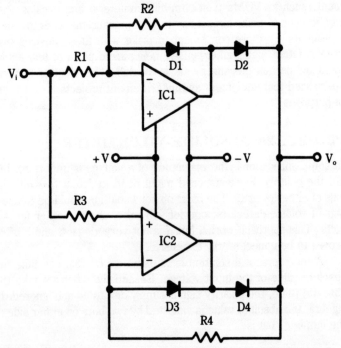

Fig. 5-1. Project 26 measures the absolute value of the input voltage.

Table 5-1. Parts List for Project 26.

Component	Part
IC1, IC2	op amp (741, or similar)
D1 – D4	1N914 diode
R1 – R4	10k, ¼-watt resistor

grade op-amp IC will not give noticeably better operation for the increased cost. In a few highly specialized and critical applications, a high-grade op-amp can be used, but it usually isn't necessary.

Note that IC1 is wired as an inverting amplifier, and IC2 is hooked up as a non-inverting amplifier. This allows the circuit to detect signals of either polarity.

As an example, let's assume that the input voltage (V_i) is positive. Under these circumstances, the output of IC1 goes negative, and also reverse biases diodes D1 and D2. The output of IC1 is blocked effectively from the circuit output (V_o). Meanwhile, IC2's output is positive. Because this op amp is designed to have unity gain, the output of IC2 is equal to V_i. Diodes D3 and D4 are forward-biased by this positive voltage, so the signal is fed out through the circuit output (V_o).

Now, if the input voltage (V_i) goes negative, just the opposite happens. IC1 now puts out a positive signal equal to the inverted value of V_i. Diodes D1 and D2 are forward-biased, which allows the signal to reach the circuit output (V_o). IC2, on the other hand, has a negative output, which reverse biases diodes D3 and D4. This signal is blocked from the circuit output (V_o).

Regardless of the polarity of the input voltage, the output (V_o) of this circuit will always be positive, with a magnitude equal to that of the input signal (V_i). Both op-amp stages have unity gain, because the input and feedback resistors have equal values.

The output from this circuit can be used to drive a dc voltmeter. You can measure an unknown voltage without worrying about the polarity of the test leads. Alternatively, the output can be used to drive any voltage-controlled circuit.

The only restriction on the input signal is that the voltage should not be allowed to exceed the IC's supply voltage.

PROJECT 27: DECIBEL METER

Most meter circuits offer linear measurements. A change in the measured voltage results in an equal amount of movement of the meter's pointer, regardless of the value. In most applications, linear measurement is highly desirable, but there are some exceptions. In some cases, measurements should be made in decibel (dB) form.

The circuit for this decibel meter project is shown in Fig. 5-2. A typical parts list for this project is given in Table 5-2. The circuitry is fairly straightforward. IC1 is a logarithmic amplifier, and IC2 is a non-inverting buffer. The output can drive a dc voltmeter. The original scale will be inaccurate, because the meter will respond to the input signal in an exponential (decibel) manner, rather than linearly.

The IC2 buffer stage is needed to prevent loading down the logarithmic amplifier stage, and to drive the voltmeter (M1).

PROJECT 28: NULL METER

In some applications, you don't really need a precise read-out of the signal voltage. You just need to know if it is equal to a specific reference level. This is true of many circuits that have to be tuned. This job can be done most easily with a null voltmeter.

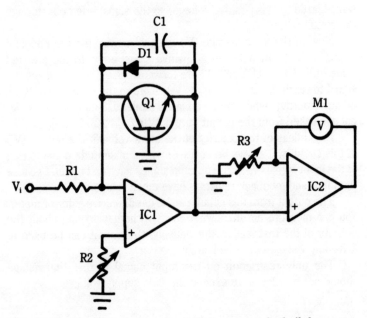

Fig. 5-2. Project 27 is a meter circuit that measures in decibels.

Table 5-2. Parts List for Project 27.

Component	Part
IC1, IC2	op amp (741, or similar)
Q1	npn transistor (2N2222, or similar)
D1	1N4001 diode
C1	0.1-μF capacitor
R1	10k, ¼-watt resistor
R2, R3	10k trimpot
M1	dc voltmeter with range to suit application

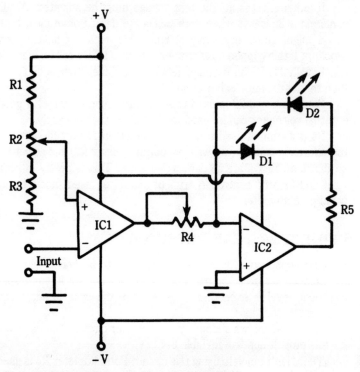

Fig. 5-3. Project 28 is a simple null-meter circuit.

Table 5-3. Parts List for Project 28.

Component	Part
IC1, IC2	op amp (741, or similar)
D1, D2	LED
R1, R3	10k, ¼-watt resistor
R2	100k potentiometer
R4	50k trimpot
R5	270-ohm, ¼-watt resistor

A *null voltmeter* is essentially a specialized comparator circuit. Two inputs are required: the voltage being tuned or tested, and a precise reference voltage. When the test voltage equals the reference voltage, the null meter indicates a null condition.

The circuit for this project appears in Fig. 5-3. Table 5-3 shows the parts list. A pair of LEDs are used as indicating devices. No expensive and bulky meter movement is needed.

If either LED is lit, the test voltage must be adjusted. A null condition is indicated when both LEDs are dark. Sometimes, both LEDs might glow very dimly at null. This does not indicate any problem. If this happens, remember that both LEDs will be more or less equally lit. If one is completely dark and the other is dimly lit, then the test voltage is not at null.

Potentiometer R2 is used to set the reference voltage. A good external voltmeter can be used to set this voltage precisely.

R4 is a trimpot used to fine-tune the circuit. Short the input to the reference voltage (the wiper of potentiometer R2) and adjust R4 so that both LEDs are as dark as possible. This is pretty much a "set and forget" control. A trimpot should be used, not a front-panel potentiometer.

PROJECT 29: PEAK DETECTOR

In some applications, you don't really need to be concerned with the exact signal level. You only need to know when (or if) some specific peak level is exceeded. This is often the case in audio and recording applications.

Figure 5-4 shows a simple peak-detector circuit. The parts list for this project appears in Table 5-4.

There isn't much to using this circuit. Potentiometer R4 is used to calibrate the circuit. Use this control to set the peak level.

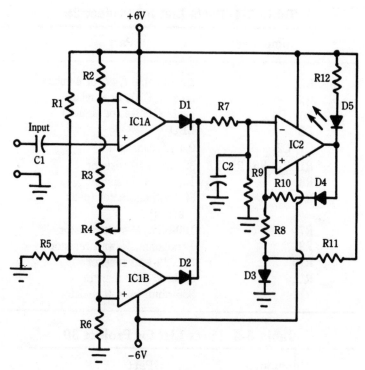

+6V

R2

R1

Input

C1

R3

R4

R5

IC1A

D1   R7

C2   R9

R10   D4

R8

D3

R12   D5

IC2

R11

IC1B   D2

R6

−6V

Fig. 5-4. Project 29 is a peak-detector circuit.

Whenever the preset peak level is exceeded, the LED lights up. If the monitored signal is below the preset peak level, the LED remains dark. That's really about all there is to this project. It's simple, but very useful, in many practical applications.

PROJECT 30: SIMPLE FREQUENCY METER

Figure 5-5 shows a very simple circuit for a frequency meter. Table 5-5 shows the parts list. This device can measure frequencies in the audible range. It can't be used for rf (radio frequency) signals.

This project uses a technique known as *time averaging* to convert the frequency of the input signal into a proportional current that can be read from the meter (M1). For maximum ease of use, replace the current meter's original scale with a customized scale. Calibrate the scale with signals of known frequencies and interpolate in-between values. If you do replace the meter scale, be very careful. The pointer and the meter winding are very delicate and

Table 5-4. Parts List for Project 29.

Component	Part
IC1	dual op amp (747, or similar)
IC2	op amp (741, or similar)
D1 – D4	diode (1N914, or similar)
D5	LED
C1, C2	0.1-μF capacitor
R1, R2, R5, R6, R8	39k, ¼-watt resistor
R3	1k, ¼-watt resistor
R4	50k potentiometer (peak adjust)
R7	100-ohm, ¼-watt resistor
R9	1-megohm, ¼-watt resistor
R10	51k, ¼-watt resistor
R11	3.9k, ¼-watt resistor
R12	330-ohm, ¼-watt resistor

Table 5-5. Parts List for Project 30.

Component	Part
IC1	LM311 amplifier
IC2	555 timer
D1	diode (1N914, or similar)
C1	560-pF capacitor
C2, C3	0.01-μF capacitor
C4	10-μF, 35-volt electrolytic capacitor
R1, R2	1k, ¼-watt resistor
R3	100k, ¼-watt resistor*
R4	56k, ¼-watt resistor
R5	250k trimpot (calibrate)
M1	current meter (0 to 50 μA)

*Experiment with other values to change the circuit's range. Decrease the resistance to read higher frequencies, or vice versa.

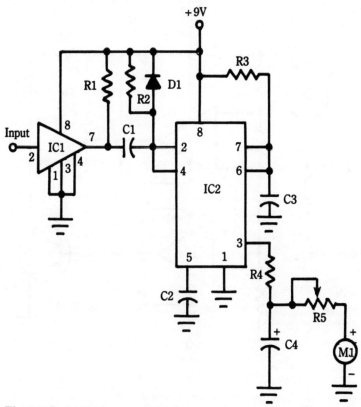

Fig. 5-5. Project 30 is a meter circuit that can measure audio frequencies.

can be damaged easily. As an alternate, you can make up a current-frequency conversion chart and mount it on the project's case. Choose a convenient location, as close to the meter as possible.

PROJECT 31: LIGHT-RANGE DETECTOR

This project will tell you if the light in a given location is within a specific range. There are many possible applications. Those applications related to photography are the most obvious, of course.

The circuit is shown in Fig. 5-6. The parts list appears in Table 5-6.

The light sensor is a simple photoresistor. The two op amps are arranged as a limit comparator. Resistors R3, R4, and R5 set the reference levels. You might want to experiment with values other than those listed in the parts list. These three resistors do not necessarily have to be equal in value. By selecting the proper resist-

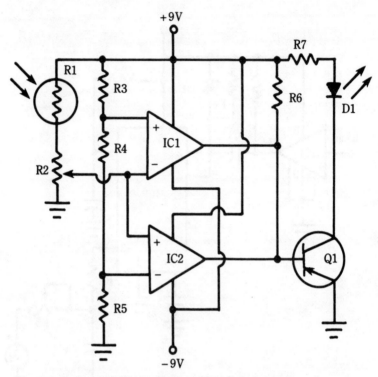

Fig. 5-6. Project 31 indicates if the ambient light is within a given range.

Table 5-6. Parts List for Project 31.

Component	Part
IC1, IC2	op amp (741, or similar)
Q1	pnp transistor (2N3906, or similar)
D1	LED
R1	photoresistor
R2	250k potentiometer
R3, R4, R5	22k, ¼-watt resistor
R6	10k, ¼-watt resistor
R7	330-ohm, ¼-watt resistor

ances, you can customize the detected range to perfectly suit your individual application. Potentiometer R2 adjusts the sensitivity of the light sensor (R1).

When the light striking the sensor is within the desired range, the LED lights up. If the light reaching the sensor is too dim or too bright, the LED turns off.

It would not be difficult to adapt this circuit to drive an external output device in place of, or together with, the LED.

PROJECT 32: TEMPERATURE-RANGE METER

There are many instances that require monitoring of temperature. In some cases a mechanical, mercury-filled thermometer is used, in others, an electronic temperature sensing device is required. Electronic thermometers are widely available now, but they tend to be fairly complex and rather expensive devices. They usually feature digital read-outs that add considerably to the complexity and expense of any electronic device. If that is what you need, then you have to accept it. Electronic digital thermometers aren't outrageously expensive, but in some applications, the digital read-out is definitely overkill. Often you don't really need to know the exact temperature, just whether or not the sensed temperature is within a specific range. This project neatly does the job, and at a minimum of cost and circuit complexity.

The schematic diagram for the temperature-range meter project appears in Fig. 5-7. Table 5-7 shows the parts list.

All semiconductors are temperature sensitive. Any pn junction reacts to changes in temperature. Therefore, you can use an ordinary diode (D10) as the temperature sensor. It isn't more than roughly accurate. It would be difficult to design a full electronic thermometer with any accuracy using such a simple sensor, but in this project you just need a rough indication of the temperature range, so a simple diode sensor does the job just fine.

Any silicon diode can be used as the temperature sensor. It must be a silicon type. Do not substitute a germanium diode.

IC1 is a LM324 quad op amp. The four sections are used as a four-stage comparator. Each stage has its own output indicator LED (D2, D4, D6, and D8). Therefore, five ranges can be indicated. The more LEDs that are lit, the higher the sensed temperature.

You might wonder how four LEDs can indicate five ranges. The fifth range is indicated if all four LEDs are dark. This means the sensed temperature is lower than the lowest indicated range.

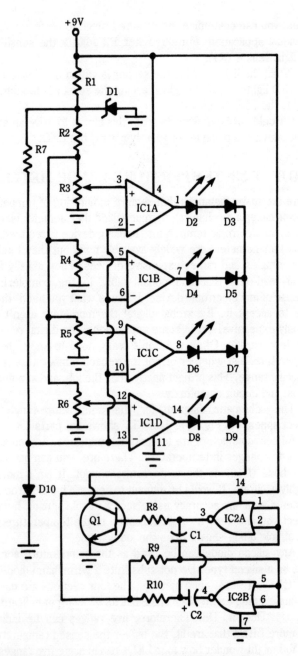

Fig. 5-7. Project 32 is a temperature-range meter circuit.

Table 5-7. Parts List for Project 32.

Component	Part
IC1	quad op amp (LM324, or similar)
IC2	CD4001 quad NOR gate
D1	6.2-volt zener diode
D2, D4, D6, D8	LED
D3, D5, D7, D9	1N914 diode
D10	silicon diode (sensor) (1N914, or similar—see text)
Q1	npn transistor (2N3904, or similar)
C1	0.82-μF capacitor
C2	0.01-μF capacitor
R1	4.7k, ¼-watt resistor
R2	47k, ¼-watt resistor
R3 – R6	50k trimpot (see text)
R7	100k, ¼-watt resistor
R8	5.1k, ¼-watt resistor
R9, R10	620k, ¼-watt resistor

LEDs are notorious current hogs. To minimize current drain, the LEDs are blinked on and off. This is accomplished by IC2, which is connected as an oscillator. Transistor Q1 is the LED driver. Diodes D3, D5, D7, and D9 prevent the four comparator stages from interfering with one another. If these diodes were not used, all four LEDs would blink on and off continuously, regardless of the sensed temperature. Obviously, that would defeat the whole purpose of this project. With these diodes in place, an LED can blink only when its associated comparator stage has a high output.

Potentiometers R3, R4, R5, and R6 are trimpots. These controls should be adjusted to give the desired range indications. If you prefer, fixed resistors could be substituted for these trimpots. In either case, what you have here is a simple resistive voltage-divider network.

6

Alarm Circuits

Unfortunately, in this day and age, security is a major concern for most of us. Burglar and safety alarms are popular products. This chapter features a number of security devices you can build yourself for a fraction of the retail cost. Although these projects are not very sophisticated, they provide adequate protection in many practical circumstances.

PROJECT 33: POWER-FAILURE DETECTOR

You've probably had the experience of coming home and finding your digital clocks blinking away. Even worse, if you have only old-fashioned dial-type electric clocks, you might not notice anything is wrong. The clocks will simply be wrong. Other electrical devices can be affected by a power failure. An obvious example is any type of computer or computer-controlled device. Any programming is lost when power is interrupted.

In some cases, you might need to know how long the power has been off. To alert you to a power-failure condition, and give you a rough indication of its length, you can use the circuit shown in Figs. 6-1 and 6-2. The parts list for this project is given in Table 6-1.

Of course, this circuit must be powered from the ac lines, or there is no point to the project at all. Use extra care in building and housing any project that uses ac power. Make sure all circuitry is well insulated, and there is no way for anyone to get a shock from handling or operating the device.

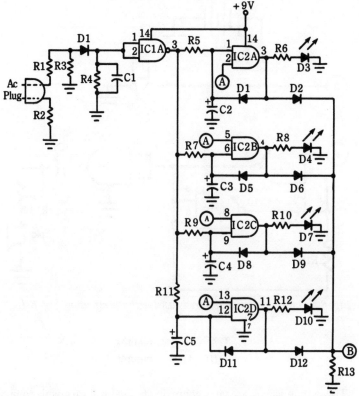

Fig. 6-1. This circuit for Project 33 indicates when a power failure has occurred.

The operation of the power-failure alarm project is very simple. Just leave the device plugged into a free ac socket. As long as the power is not interrupted, the circuit does nothing at all. You can just forget about it and go on about your business.

Whenever ac power is interrupted, the circuit sounds a tone. Moreover, by looking at the number of lit LEDs, you can get a rough idea of how long the power was out. When power is restored, the alarm will continue to sound and the LEDs will remain lit.

Using the component values given in the parts list, the time period for each of the LEDs is as follows:

D3	2 seconds
D4	10 seconds

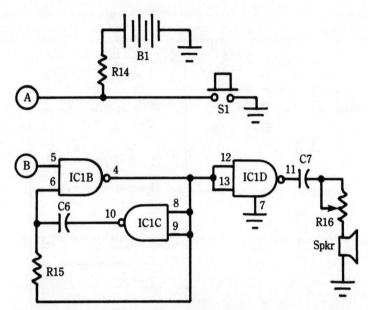

Fig. 6-2. This auxiliary circuitry is used with the circuit of Fig. 6-1.

| D7 | 1½ minutes |
| D10 | 8 minutes |

These time values are very approximate, and will vary with component tolerances.

You can set up other timing periods by changing the value(s) of the appropriate capacitor(s):

D3	C2
D4	C3
D7	C4
D10	C5

B1 must be a nine-volt battery. Don't use an ac-to-dc power supply. The circuit must receive power when ac power is interrupted. The current drain on this battery is very low. Unless you have a lot of power failures, a battery should last several months, perhaps even a year. There is an easy way to check the battery. Just simulate a power failure by unplugging the alarm. It should go off.

Table 6-1. Parts List for Project 33.

Component	Part
IC1	CD4011 quad NAND gate
IC2	CD4081 quad AND gate
D1, D2, D5, D6, D8, D9, D11, D12	diode (1N914, or similar)
D3, D4, D7, D10	LED
C1	0.1-μF capacitor
C2	2.2-μF electrolytic capacitor*
C3	10-μF electrolytic capacitor*
C4	100-μF electrolytic capacitor*
C5	470-μF electrolytic capacitor*
C6	0.022-μF capacitor**
R1, R2	2.2-megohm resistor
R3	470k resistor
R4, R6, R8, R10, R12	10k resistor
R5, R7, R9, R11	1-megohm resistor
R13, R14	100k resistor
R15	33k resistor**
S1	normally open SPST pushbutton switch

*timing capacitor—see text
**frequency determining component—experiment with other component values to change alarm frequency.

When the alarm has gone off, it can be reset easily. Just briefly close switch S1. The alarm tone will be cut off, and any lit LEDs will be turned off.

For maximum clarity, the schematic is shown in two separate figures. Figure 6-1 shows the monitor and LED control circuitry. Figure 6-2 shows the reset and alarm tone circuits. All points marked A in Fig. 6-1 should be connected to point A of the reset circuit (Fig. 6-2). Similarly, point B in Fig. 6-1 should be connected to point B of the alarm tone generator in Fig. 6-2.

PROJECT 34: ELECTRONIC SECURITY LOCK

Combination locks can be tough to beat, but a patient thief with good ears and a good sense of touch usually can open any mechanical combination lock. An electronic combination lock can be tougher to crack because the circuit can be designed with a time limit to enter the correct combination. An incorrect entry can disable the circuit for a period of time. It would require a very, very patient thief to try all possible combinations.

An electronic combination lock can be used in place of a mechanical lock. An output relay controls one or more solenoids that perform the actual locking operations. The solenoids don't move until the relay is activated.

Another application for an electronic combination lock is to restrict access to certain electronic equipment. The equipment will not work unless the proper combination code is entered.

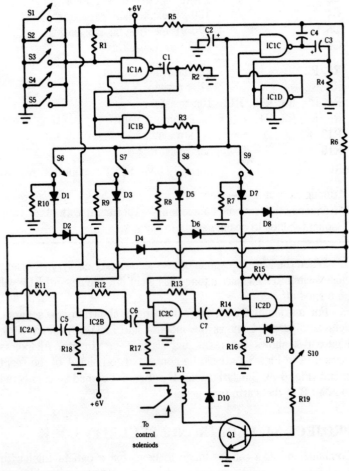

Fig. 6-3. Project 34 is an electronic combination lock.

Table 6-2. Parts List for Project 34.

Component	Part
IC1	CD4011 quad NAND gate
IC2	CD4081 quad AND gate
Q1	npn transistor (2N2222, or similiar—see text)
D1 – D10	diode (1N914, or similar)
C1	50-μF, 35-volt electrolytic capacitor
C2	1-μF, 35-volt electrolytic capacitor
C3, C5, C6, C7	10-μF, 35-volt electrolytic capacitor
C4	0.01-μF capacitor
R1, R14, R19	10k, ¼-watt resistor
R2	2.2-megohm, ¼-watt resistor
R3, R15, R11, R12, R13, R15	100k, ¼-watt resistor
R4	1.5-megohm, ¼-watt resistor
R6	1k, ¼-watt resistor
R7, R8, R9, R10	4.7k, ¼-watt resistor
R16, R17, R18	220k, ¼-watt resistor
K1	relay to suit load
S1 – S10	normally open SPST swithces (10-switch keypad)

The electronic combination lock circuit shown in Fig. 6-3 provides very good security. There are 10,000 possible combinations. With the component values listed in the parts list of Table 6-2, the correct combination must be entered within a seven-second period.

The combination is entered by pressing buttons on a ten-switch keypad. If an incorrect key is pressed, the keypad will be disabled. In addition, an external alarm could be sounded.

To operate the lock, the following switches must be pressed in sequence:

S6
S7
S8

Switches S1 through S5 are dummies. If any one of these switches is closed, the shut-down/alarm process activates. Closing any of the correct combination switches out of sequence also disables the lock.

You can set up any combination you like by connecting the switch leads to the appropriate positions of the keypad, that is, the switches are not arranged in numerical order. Pick a combination that has some relevance to you, so it will be easy to remember. You could base the combination on your birthdate, or that of a loved one, or perhaps a phone number.

Transistor Q1 should be selected to supply sufficient current for the desired application. It depends on what you want the lock to control. In some applications, where the electronic lock is restricting control of a piece of equipment, the relay (K1) can be eliminated. Of course, diode D10 should also be eliminated in that case. The function of this diode is simply to protect the relay coil from back-emf.

PROJECT 35: BURGLAR-ALARM
CONTROL CIRCUIT

The circuit shown in Fig. 6-4 can be used as the heart of a complete burglar-alarm system. When the protected area is violated, relay K1 is activated, sounding an alarm, siren, or other device. The relay (K1) and protective diode (D6) can be eliminated in some applications.

Switches S1, S2, and S5 are normally open type switch devices. If any of these switches are closed, even briefly, the alarm sounds. Switches S3 and S4 are normally closed devices. Opening one of the normally closed switches also triggers the alarm.

Switch S6 (also a normally open unit) is the system reset switch. The alarm can be turned off by briefly closing this switch. The reset switch should be hidden, so an intruder won't be able to disable the alarm himself.

Depending on what is being driven by this control circuit, you might need to replace transistor Q1 with a unit that has a higher current rating than the transistor listed in Table 6-3.

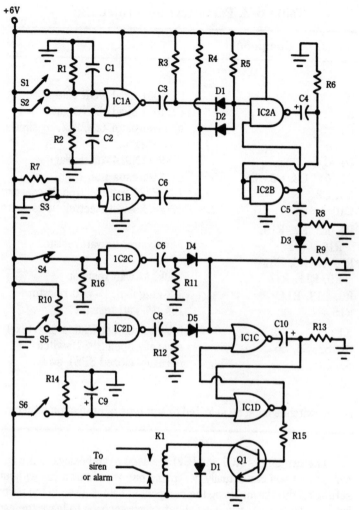

Fig. 6-4. Project 35 can be used as the control center of a home security system.

PROJECT 36: AUTO ALARM

This project is a simple alarm system for your car. When the circuit is activated, anyone opening any of the car doors triggers the alarm.

Switch S2 is used to arm or disarm the alarm system. You could use a lock switch outside the car, or S2 could be placed inside the vehicle. A lock switch is still advisable.

Table 6-3. Parts List for Project 35.

Component	Part
IC1	CD4001 quad NOR gate
IC2	CD4011 quad NAND gate
Q1	npn transistor (2N2222, or similiar —see text)
D1 – D6	diode (1N914, or similar)
C1, C2, C3, C5 – C8	0.01-μF capacitor
C4, C9	10-μF, 25-volt electrolytic capacitor
C10	47-μF, 25-volt electrolytic capacitor
R1, R2, R7, R10, R16	1-megohm, ¼-watt resistor
R3, R4, R5, R8, R9, R11, R12	100k, ¼-watt resistor
R6, R13, R14	3.3-megohm, ¼-watt resistor
R15	10k, ¼-watt resistor
K1	relay to suit application (see text)
S1, S2, S5, S6	normally open SPST switch
S3, S4	normally closed SPST switch

Note: switches must be selected to suit type of protection required.

The two timers (IC1 and IC2) are used to add delays to the circuit so that you can actually get in and out of your own car without setting off the alarm. Using the component values given in the parts list (see Table 6-4), you have about seven seconds to leave the car and close the door after arming S2. When you open the door to get back in, you have about five seconds to disarm S2. You can increase either delay time by increasing the value of the appropriate capacitor (C1 for arming, C4 for disarming). Of course, decreasing the capacitance reduces the delay time.

Figure 6-5 shows the schematic for this project. The circuit can be powered by your car battery. It does not represent a significant current drain. The car's horn can be used as the alarm signalling device.

Table 6-4. Parts List for Project 36.

Component	Part
IC1, IC2	555 timer
Q1	pnp transistor (2N4403, or similar)
D1	diode (1N4002, or similar)
C1	15-μF, 25-volt electrolytic capacitor
C2, C4	0.01-μF capacitor
C3	0.01-μF capacitor
C5	10-k, μF, 25-volt electrolytic capacitor
R1	470k, ¼-watt resistor
R2	270-ohm, ¼-watt resistor
R3	2.2-megohm, ¼-watt resistor
S1	SPST switch (ignition switch)
S2	SPST switch (lock switch preferred)
K1	relay (contacts to suit load)

PROJECT 37: MOVEMENT ALARM

This project can be used to protect almost any object. When the object is moved, the alarm goes off.

Figure 6-6 shows the circuit for this project. The parts list appears in Table 6-5.

This project can offer two different types of protection. Any normally closed switch or similar device can be used. This includes the foil tape used in many burglar-alarm systems. When the foil is broken, it is the same as opening a normally closed switch.

The other type of protection is to use a vibration sensor. This should also be a normally closed type. These devices are usually sold at stores that carry alarm-related equipment. Any motion or vibration causes the sensor's switch to open, setting off the alarm. The vibration sensor can be mounted physically on almost any object you wish to protect. You can even protect any sensitive

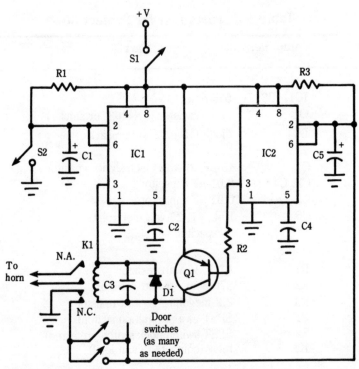

Fig. 6-5. Project 36 is an alarm circuit that can protect your car.

Component	Part
IC1, IC2	CD4011 quad NAND gate
Q1	npn transistor (2N2222, or similiar—current rating must be sufficient for desired load)
D1	diode (1N914, or similiar)
D2	diode (1N4002, or similiar)
C1	33-μF, 35-volt electrolytic capacitor
C2	25-μF, 35-volt electrolytic capacitor
C3	500-pF capacitor
R1	6.8k, 1/4-watt resistor
R2, R4	1-megohm, 1/4-watt resistor
R3	10-megohm, 1/4-watt resistor
R5	220k, 1/4-watt resistor
R6	27k, 1/4-watt resistor
S1	SPST switch—lock switch preferred
S2, S3	normally closed sensor (see text)

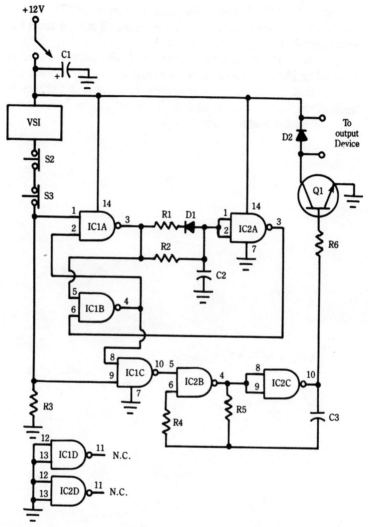

Fig. 6-6. This alarm circuit for Project 37 can be used to protect almost anything.

papers you don't want unauthorized persons to look at. Just use the vibration sensor as a paperweight. Nobody will be able to get at the papers without moving the sensor, and the sensor cannot be moved without setting off the alarm.

System arming switch S1 should be hidden, or you can use a lock switch, which requires a key. One neat little trick is to mount several identical-looking switches. One of these is S1. The others are closed switches in parallel with the sensor. If anyone flicks the wrong switch, the alarm will be set off.

A momentary break in the sensor circuit also causes the alarm to sound for about 20 seconds. If the sensor device is reclosed, the alarm circuit resets itself automatically. If there is a permanent break in the sensor circuit, such as if the foil tape is broken, the alarm will sound continuously.

7

Miscellaneous Projects

This final chapter is a sort of hodge-podge of useful and interesting projects that don't quite fit into any of the other chapter headings. Although these projects don't fit into a neat classification, you should still find them worthwhile.

PROJECT 38: FREQUENCY DIVIDER

In some applications, it is necessary to reduce a signal frequency. This project does the trick in many cases.

Figure 7-1 shows the schematic for this project. The parts list appears in Table 7-1.

This circuit accepts a high-frequency input, and divides the frequency at the output by any integer from 1 to 10 via rotary switch S1.

Because this circuit is built around digital ICs, it works best if the input is in the form of rectangle waves. Other waveshapes at the input could produce erratic operation. You might want to precede this circuit with a Schmitt trigger to square off analog waveforms. The output will be a square wave.

As an example of how this circuit functions, let's assume that the input signal has a frequency of 3500 Hz (3.5 kHz). The output frequency for each position of S1 will be as follows:

$$
\begin{array}{llll}
1 & F/1 & = & 3500 \text{ Hz} \\
2 & F/2 & = & 1750 \text{ Hz}
\end{array}
$$

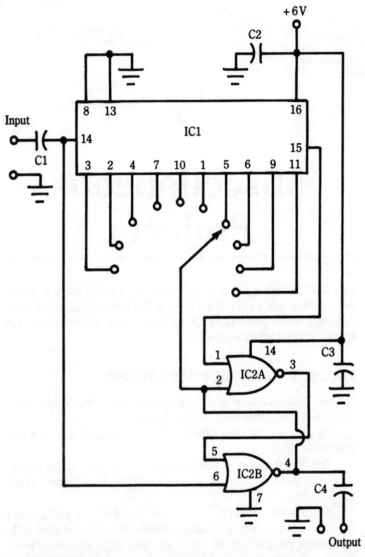

Fig. 7-1. This circuit for Project 38 can divide an input frequency by any integer from 1 to 10.

3	F/3	=	1166.7 Hz
4	F/4	=	875 Hz
5	F/5	=	700 Hz
6	F/6	=	583.3 Hz
7	F/7	=	500 Hz

Table 7-1. Parts List for Project 38.

Component	Part
IC	CD4017 decade counter
IC2	CD4001 quad NOR gate
C1, C4	$0.1 - \mu F$ capacitor
C2, C3	$0.01\text{-}\mu F$ capacitor
S1	1-pole, 10-throw rotary switch

8	F/8	=	437.5 Hz
9	F/9	=	388.9 Hz
10	F/10	=	350 Hz

In an electronic music system, some very interesting effects can be achieved by mixing the original signal with one of the divided values.

PROJECT 39: FREQUENCY MULTIPLIER

This project is just the opposite of the previous one. This time the input is a low-frequency signal, and the output is at a higher frequency. Once again, because digital ICs are used, the circuit should be used with rectangle waves.

The frequency multiplier circuit is illustrated in Fig. 7-2. The parts list for this project is given in Table 7-2. This is a very simple project. It calls for just two identical ICs and two identical resistors. The resistance value is not even very critical.

This frequency-multiplier project offers two independent outputs. Output A multiplies the input frequency by a factor of two. The second stage is identical to the first. Therefore, the signal at output B is four times the frequency of the input signal (or twice the frequency of output A).

In musical terms, output A raises the pitch one octave, and output B raises the pitch two octaves over its original value. The two outputs can be used separately, or together.

92

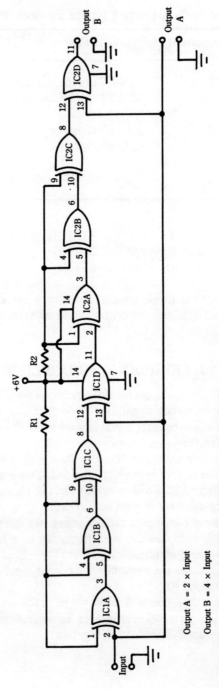

Fig. 7-2. Project 39 multiplies the input frequency by two and by four.

Output A = 2 × Input

Output B = 4 × Input

Table 7-2. Parts List for Project 39.

Component	Part
IC1, IC2	74C86 quad exclusive-OR gate
R1, R2	2.2k, ¼-watt resistor

PROJECT 40: DIGITAL FILTER

I suppose you could say this project is somewhat related to the frequency divider and frequency-multiplier projects presented earlier. A filter is a frequency-sensitive circuit. It permits some frequencies to pass through to the output, while other frequencies are blocked.

Filtering is basically an analog function, but this project uses digital circuitry as a filter. Analog signals can be filtered with this circuit.

The schematic for the digital-filter project appears in Fig. 7-3. Table 7-3 shows a typical parts list for this project.

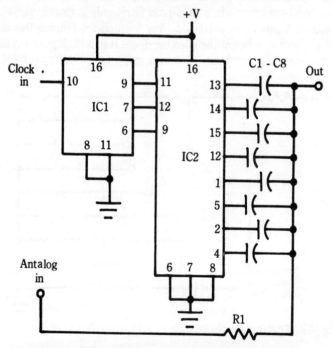

Fig. 7-3. Project 40 is a digital circuit that performs analog filtering.

Table 7-3. Parts List for Project 40.

Component	Part
IC1	CD4040 BCD ripple counter
IC2	CD4051 BCD-to-decimal decoder
C1 – C8	0.01-μF capacitor*
R1	1k, ¼-watt resistor*

*Experiment with other values. Note that all capacitors must be equal

In technical terms, this circuit is a *commuting filter*. A very simplified form of the commuting filter appears in Fig. 7-4. A number of identical capacitors are sequentially switched in and out of the circuit. If just one capacitor was used, this would be a low-pass filter, but the capacitor-switching action changes the operation of the filter to a band-pass type. In a band-pass filter, only a specific range, or band, of frequencies is passed. Any frequencies outside this band (either above or below the pass band) are rejected. Figure 7-5 illustrates the basic frequency-response graph for a band-pass filter.

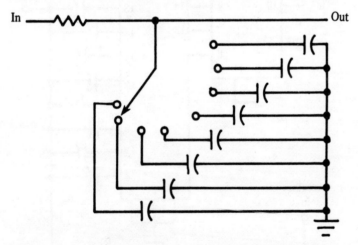

Fig. 7-4. This is a simplified form of the commuting filter.

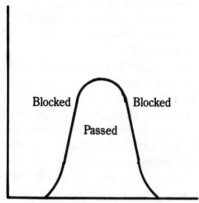

Fig. 7-5. This is the frequency-response graph for a typical band-pass filter.

For this commuting filter the center frequency of the pass band can be found with the following formula:

$$F_c = 1/2nRC$$

where
    F_c is the center frequency
    R is the resistor value
    n is the number of capacitors in the switching sequence and
    C is the value of one of the capacitors

All of the capacitors should have equal values for this type of circuit to work properly.

Actually, a practical commuting filter has multiple passbands. Harmonics (integer multiples) of the nominal center frequency (as defined by the above formula) also pass through to the output. For instance, if $F_c = 1000$ Hz, there will be additional passbands centered around the following frequencies:

2000 Hz	(second harmonic)
3000 Hz	(third harmonic)
4000 Hz	(fourth harmonic)
5000 Hz	(fifth harmonic)

and so forth.

Each successive harmonic passband has a lower amplitude than its predecessor. The upper harmonics are filtered effectively out of the signal.

Figure 7-6 shows the frequency response graph for a typical commuting filter. Because the multiple passbands on this graph somewhat resemble the teeth of a comb, the commuting filter is sometimes called a *comb filter*.

If the application requires more traditional bandpass operation, a low-pass filter at the output of the commuting filter can help get rid of most of the harmonic passbands.

Return to the circuit diagram of Fig. 7-3. Note that two input signals are required—a (digital) clock to drive the counter (IC1)—and the signal to be filtered. The clock signal should be a square wave. The input signal can be any analog waveform.

The capacitor values are fairly critical in this application. Use high-grade capacitors with no more than a 10 percent tolerance rating. The component values given in the parts list are typical. Experiment with other values. Remember, all eight capacitors must have equal values.

The only real restriction on the input signal is that the peak-to-peak voltage must be less than the supply voltage used to power the two ICs. The chips could be damaged if this value is exceeded.

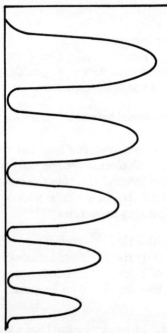

Fig. 7-6. This is the frequency-response graph for a typical commuting (or comb) filter.

PROJECT 41: RANDOM-VOLTAGE GENERATOR

The MK50240 top-octave generator IC normally is used to produce audible frequencies. A high-frequency clock signal is fed into the MK50240. This clock frequency is divided by specific values to create thirteen lower frequencies, corresponding to the notes within an octave.

If the clock frequency is very low (a few hundred hertz, or less), the divided output frequencies will be below the audible range. Combining two or more of the MK50240's outputs, a random voltage pattern will be generated.

Figure 7-7 shows a random-voltage generator circuit. The parts list for this project appears in Table 7-4.

Random voltages can be used in games and automated electronic music systems. Any voltage-controlled circuit (such as a VCO, or a VCA) can be driven randomly by this circuit.

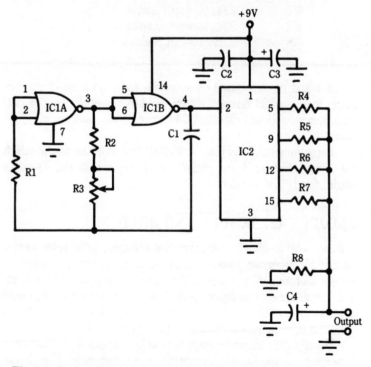

Fig. 7-7. Project 41 generates a random voltage signal.

Table 7-4. Parts List for Project 41.

Component	Part
IC1	CD4001 quad NOR gate
IC2	MK50240 top-octave generator
C1	50-μF, 25-volt electrolytic capacitor
C2	0.01-μF capacitor
C3	10-μF, 25-volt electrolytic capacitor
C4	33-μF, 25-volt electrolytic capacitor
R1	10k, ¼-watt resistor
R2	4.7k, ¼-watt resistor
R3	10k potentiometer
R4, R6	27k, ¼-watt resistor
R5	22k, ¼-watt resistor
R7	33k, ¼-watt resistor
R8	3.3k, ¼-watt resistor

IC1 and its associated components make up a low-frequency clock. Resistors R4 through R7 are used to weight the various outputs. Resistor R8 and capacitor C4 form a simple, passive, low-pass filter.

Potentiometer R3 is used to control manually the rate at which the voltage changes. If you prefer, you can replace R2 and R3 with a single, fixed, 10k resistor.

PROJECT 42: SAMPLE AND HOLD

Another way to obtain pseudorandom voltages, or in some cases, predictable recurring patterns, is to use a type of circuit known as a *sample and hold*. The name of this circuit pretty much says it all. The circuit accepts an input signal. The instantaneous voltage level is sampled, and then held at the output. Sample-and-hold circuits often are used in analog electronic-music synthesizers.

Usually the sampled voltage is held in a capacitor. No capacitor is perfect, so the sample cannot be held indefinitely. The charge gradually leaks off, as illustrated in Fig. 7-8. This generally isn't much of a problem, because in most practical applications, the input signal is sampled repeatedly at a regular rate (determined by the frequency of a clock signal). Before the charge can leak off the capacitor, a new sample is taken.

Figure 7-9 illustrates the operation of a sample and hold. In this case, a random-noise signal is used as the input, but any varying voltage can be used. (Obviously, there isn't much point in sampling a constant dc voltage.) When a noise signal is used as the input signal, the output is aperiodic. There is no repeating pattern. (If the noise source is pseudorandom, the pattern does eventually repeat, but it is very long.)

If the input signal is a periodic ac waveform, such as a sine wave or a ramp wave, and the signal frequency is harmonically related to the clock frequency, then a repeating pattern will occur at the output.

Generally, the clock frequency is quite low in comparison with the signal frequency. In most practical applications, the clock frequency should be below the audible range. However, some very unusual modulation effects sometimes can be achieved by using a clock frequency in the audible range. The output is an audible tone with lots of sidebands, and it sounds rather raspy and buzzy. The effect probably won't be too pleasant without some external modification to the signal (filtering, and so forth).

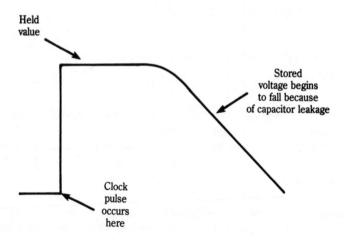

Fig. 7-8. The charge will gradually leak from a practical capacitor.

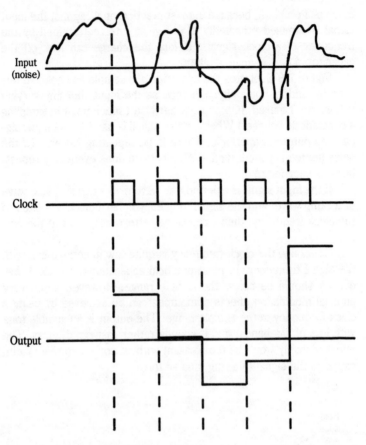

Fig. 7-9. If the input is random noise, the output is a series of random dc voltages.

The schematic diagram for this sample-and-hold project appears in Fig. 7-10. Table 7-5 shows the parts list.

The actual sample and hold is built around an op amp (IC1). In this application, it is strongly advisable to use a high-quality, low-noise op-amp chip such as the 308 or the 536, among others.

The op amp is connected as a buffer amplifier with a gain of 1 (unity). Its function is to help minimize leakage into the circuit connected to the output. Basically, the op-amp's job is to isolate the storage capacitor (C3) from the load.

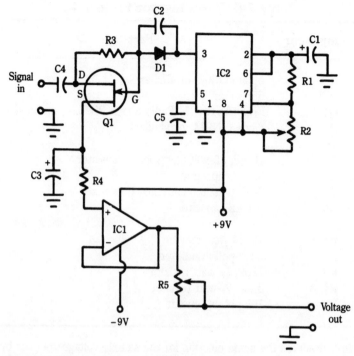

Fig. 7-10. Project 42 is a practical sample-and-hold circuit.

FET Q1 and capacitor C3 really do most of the work in this sample-and-hold circuit. A negative-going pulse on Q1's gate causes C3 to be discharged quickly and then charged back up to the new sample level. Capacitor C3 must be a tantalum type. Electrolytic capacitors have far too much leakage for this application, even for experimental purposes. If you substitute an electrolytic capacitor, the circuit will be virtually useless.

The storage-capacitor's value must be small enough so that it can charge and discharge quickly, yet high enough so that it will hold the sample voltage an adequate length of time. The 5-μF capacitor specified in the parts list seems to work fairly well.

IC2, along with its associated components, forms an astable multivibrator, or rectangle-wave generator. The frequency can be adjusted manually via potentiometer R2. This signal generator provides the clock signal for the actual sample-and-hold circuitry.

Potentiometer R5 is a master output-level control. If the output were an audio signal, we would call this potentiometer a *volume con-*

Table 7-5. Parts List for Project 42.

Component	Part
IC1	op amp (308, or similar)
IC2	555 timer
Q1	FET (A5T3821, Radio Shack RS2028, or similiar)
D1	diode (1N914, or similar)
C1	10-μF, 25-volt electrolytic capacitor
C2	50-pF capacitor
C3	5-μF tantalum capacitor (see text)
C4	0.1-μF capacitor
R1	1k, ¼-watt resistor
R2	500k potentiometer
R3	120k, ¼-watt resistor
R4	2.2k, ¼-watt resistor
R5	10k potentiometer

trol. It serves the same function for the sample voltages put out by this circuit.

Typically, the output from a sample and hold is used as a control signal for some voltage-controlled device (or devices), such as a VCA (voltage-controlled amplifier) or VCO (voltage-controlled oscillator).

PROJECT 43: PULSE-WIDTH MODULATOR

In recent years you may have heard of pulse-width modulation, which is usually abbreviated as PWM. This project will let you use PWM techniques yourself.

Modulation is a process of imposing a program signal onto a second carrier signal for the purposes of transmission or storage. The two most familiar types of modulation are am (amplitude modulation) and fm (frequency modulation). In *amplitude modulation*, the amplitude, or instantaneous level of the carrier signal, is varied in step with the program signal. In *frequency modulation*, the frequency of the carrier signal is raised or lowered in response to variations in the program signal. With this in mind, it shouldn't be too hard to figure out what pulse-width modulation does.

In an ordinary rectangle-wave signal, all of the pulses are the same, as illustrated in Fig. 7-11. Each pulse lasts exactly as long as the previous one and the one after it. A PWM system uses such a rectangle wave as the carrier.

As in any modulation system, two inputs are required, as shown in Fig. 7-12. In addition to the carrier signal (V_c), There is a program or modulating signal (V_m). These two signals are combined at the output (V_o).

At any given moment, the instantaneous amplitude of the V_m signal determines the width of the current output pulse. The output is a string of rectangle waves with constantly changing duty cycles, as shown in Fig. 7-13. The duty cycles vary in step with the V_m (program) signal.

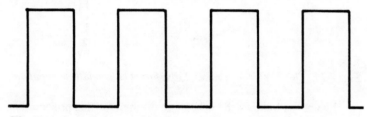

Fig. 7-11. In ordinary rectangle waves, all of the pulses are of the same width.

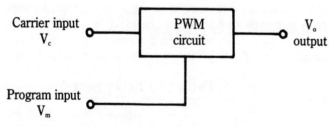

Fig. 7-12. A pulse-wave modulator requires two inputs.

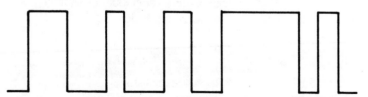

Fig. 7-13. In a pulse-wave modulator, the duty cycle of the carrier signal varies with the program signal.

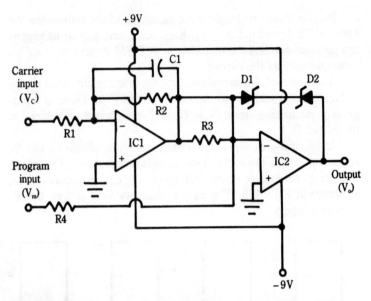

Fig. 7-14. This circuit for Project 43 performs pulse-width modulation.

Figure 7-14 shows the schematic diagram for this project. The parts list is given in Table 7-6.

Using PWM, data or an audio program can be transmitted or stored using either analog or, more commonly, digital equipment.

Of course, there is no point to any form of modulation unless there is some way to demodulate the signal at a different location and/or time to recover the original program signal. Because the amplitude of the PWM pulses remains constant, demodulation is

Table 7-6. Parts List for Project 43.

Component	Part
IC1, IC2	high-grade op amp (308, or similar)
D1, D2	6.8-volt zener diode
C1	0.01-μF capacitor
R1	27k, ¼-watt resistor
R2	1-megohm, ¼-watt resistor
R3, R4	10k, ¼-watt resistor

not very difficult. Simple low-pass filtering can produce an output voltage that is proportional to the pulse width. Because the pulse width corresponds to the program signal, the output of the low-pass filter resembles the original program signal (V_m).

For best results, the carrier signal should have a fairly high frequency. As a rough rule of thumb, the carrier frequency should be at least twice the maximum program frequency. For better fidelity, the carrier frequency should be even higher than this.

Be aware that this project performs a very crude form of PWM, so don't expect miracles. An audio recording made with this device will not come close to rivaling a CD (compact disc), even though CDs are recorded via a system that is not very different from the one described here, there are some differences. This modulator project is basically for experimental use only.

Of course, the quality of the signal depends on the type of chips you use. The higher the grade of op amps used, the cleaner the modulated signal will be.

PROJECT 44: SLAVE FLASH CONTROLLER

Flash strobe lights often are used in photography. This project uses a trigger pulse to fire almost any flash whenever desired. The flash can be very brief, effectively "freezing" the action.

Table 7-7. Parts List for Project 44.

Component	Part
IC1	CD4011 quad NAND gate
IC2	CD4017 decade counter/divider
Q1	npn transistor (2N2222, or similar)
D1, D2	diode (1N914, or similar)
C1	0.01-μF capacitor
R1	photoresistor
R2	100k potentiometer
R3, R6, R8, R9	100k, 1/4-watt resistor
R4, R5	1-megohm, 1/4-watt resistor
R7	500k potentiometer
R10	1k, 1/2-watt resistor—repeat as needed(see text)

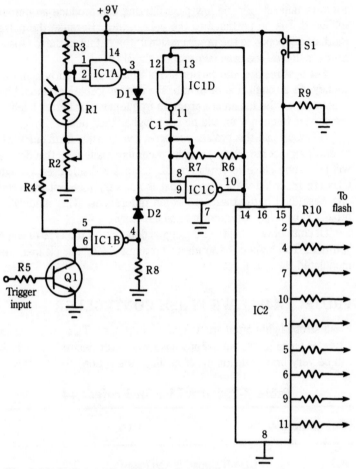

Fig. 7-15. Project 44 is a slave flash controller.

Figure 7-15 shows the circuit. The parts list for this project appears in Table 7-7.

The output of this circuit is fed to the controller input of your flash unit. In some cases, you might need to add an SCR, as illustrated in Fig. 7-16. It depends on what kind of firing system (if any) your flash unit has built into it.

An external trigger signal can be used to fire the flash. Use the input connection in Fig. 7-15. Alternatively, the unit also can function as an automated "slave" flash. Light from the main flash is sensed by the photoresistor (R1). When the light striking this sensor exceeds a specific level, the circuit fires the remote flash. Potentiometer R2 is used to adjust the sensitivity of the sensor.

The CD4017's multiple outputs can be used to drive multiple firing circuits (the same flash unit often can be used repeatedly. The driving resistor (R10) and any other firing circuit (such as an SCR, if used) just needs to be repeated for each counter output. The timing of these successive flashes can be controlled by potentiometer R7. The sequence can be halted and the counter reset at any time simply by momentarily closing switch S1. The counter also must be manually reset after it has stepped through a complete output sequence. If just a single output is used, you will need to press the reset button after each flash. This can be a minor nuisance in some cases, but it could be quite desirable in others. The limitations of the design are intended to keep the project cost down. The complete circuit as shown in Fig. 7-15 can be built for under $10.

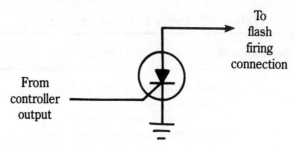

Fig. 7-16. Some flash units might require an SCR to fire.

PROJECT 45: SINGLE-TO-DUAL-POLARITY CONVERTER

Many electronics circuits (especially digital circuits) demand regulated power supplies. Often a dual polarity capability (both positive and negative voltages) is required.

This project accepts an unregulated single-polarity dc input voltage and puts out regulated dual-supply voltages. Remember, the input voltage must supply current to both the positive (V+) and negative (V–) outputs, in addition to the current consumed by the regulation process.

Figure 7-17 shows the circuit for this project. The parts list appears in Table 7-8. Note that the two outputs are referenced to an artificial (floating) ground. As with any single-to-dual-polarity converter circuit, the output common should never be shorted to true

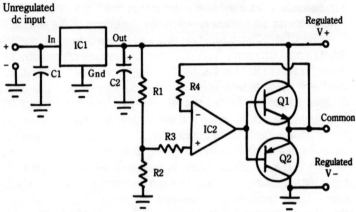

Fig. 7-17. Project 45 converts a single-polarity dc voltage into a dual-polarity voltage source.

Table 7-8. Parts List for Project 45.

Component	Part
IC1	voltage regulator—to suit application*
IC2	op amp
Q1	npn complementary transistor*
Q2	pnp complementary transistor*
C1	0.1-μF capacitor
C2	500-μF electrolytic capacitor (working voltage should exceed the input voltage by at least 20%)
R1, R2, R3, R4	10k, ½-watt resistor
*See text	

earth ground. Damage will result if you connect the common line to a true ground point. Be careful.

The conversion circuitry requires a regulated voltage. Therefore, a voltage regulator (IC1, C1, and C2) is included in the circuit. If the input voltage is already regulated, this regulator will be unnecessary, but it won't do any harm. Because the input voltage is regulated, both output voltages are regulated automatically. Only the single voltage regulator is required.

Transistors Q1 and Q2 are a pair of complementary (opposite polarity, but with equal electrical characteristics) shunt-regulator transistors. They create an artificial, or floating, ground to be used by the load circuit. Incidentally, this artificial ground is sometimes called a *counterpoise ground*.

The parts list assumes that the output load draws only a small current. If a larger output current is required, you should replace Q1 and Q2 with heftier transistors.

Resistor R4 ties the inverting input of the op amp to the common (artificial ground) point. The non-inverting input is fed a reference voltage-divider network made up of resistors R1 and R2. If the floating ground shifts for any reason, the op amp's output voltage changes by a proportional amount. If the op amp's output should happen to go negative, for example, then Q2 (pnp) starts to conduct more heavily. Conversely, if the op amp's output becomes more positive, then Q1 (npn) starts to conduct more heavily. In either case, the transistors automatically correct any shift of the common level.

When the output is balanced correctly, the two transistors should conduct equally. The circuit always tries to remain in this balanced condition.

The parts list might appear a little incomplete. Specific transistors are not listed. They should be selected to handle more than the maximum current that will ever be demanded by the load. Similarly, IC1 should be selected for the desired voltage. The regulator should be rated for a voltage that is twice the desired output voltages. That is, if a 12-volt regulator is used, the output will be ±6 volts.

You can experiment with other resistance values. For the dual polarity output voltages to be symmetrical $(V+ = V-)$ the following conditions must be met:

$$R1 = R2$$
$$R3 = R4$$

Symmetrical outputs will be desired for most applications, but you might run across an occasional exception to this rule of thumb.

PROJECT 46: DUAL-POLARITY VOLTAGE REGULATOR

If your specific application requires a true ground between the dual polarity supply voltages, you will need a true dual-polarity voltage source, of course.

If the dual supply voltages must be regulated, you could use separate positive and negative regulators, but that is a bit inefficient. This project is a dual-polarity voltage regulator. The two halves of the circuit share as many components as possible, thus minimizing the cost, circuit complexity, and overall size.

The circuit is illustrated in Fig. 7-18. Table 7-9 shows the parts list. A dual-polarity, unregulated supply serves as the input. The output is a regulated dual-polarity voltage source referenced to true earth ground.

To understand how this circuit works, let's break it up into its positive and negative sections. The positive portion of the regulator is shown separately in Fig. 7-19. This section of the circuit includes the following components: IC1, Q1, D1, R1, R2, and R3. Transistor Q1 is an npn series-pass element. It should be selected to supply sufficient current to the load. Remember to allow for a little excess current to make up for the power consumed in the regulator circuit. The base current of this transistor is controlled directly by the output of the op amp (IC1).

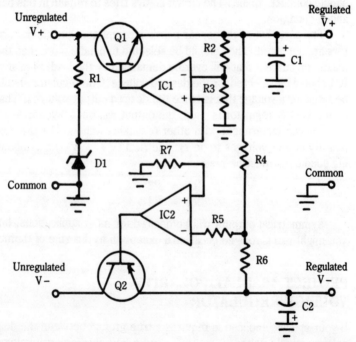

Fig. 7-18. This circuit for Project 46 is a dual-polarity voltage regulator.

Table 7-9. Parts List for Project 46.

Component	Part
IC1, IC2	op amp
Q1	npn transistor*
Q2	pnp transistor*
D1	zener diode to suit application*
C1, C2	500-μF electrolytic capacitor voltage to suit output
R1	value to suit application (see text)
R2 – R7	10k, ½-watt resistor

The non-inverting input signal for this op amp is a reference voltage determined by resistor R1 and zener diode D1. These components are selected specifically for the intended application. A portion of the output voltage is fed back to the op-amp's inverting input through the voltage divider made up of resistors R2 and R3.

When the inverting input is fed the same voltage as the non-inverting voltage, the op amp is in a stable condition, holding the output constant. If the output voltage increases or decreases for any reason, the voltage at the inverting input changes proportionately. It no longer is equal to the reference voltage at the non-inverting input, which remains constant. When this happens, the op amp's output signal changes, altering the base current of the tran-

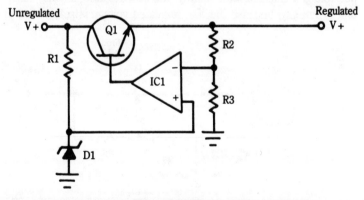

Fig. 7-19. This is the positive portion of the circuit of Fig. 7-18.

sistor. The transistor's emitter voltage (the output) is brought back into line with the nominal output voltage. The output voltage is regulated and effectively held at a constant level no matter what happens.

Even a very small change in the output voltage causes this circuit to respond very quickly to correct the error. Output ripple is negligible. For most purposes, it can be considered virtually nonexistent.

Figure 7-20 shows the negative half of the circuit. It is very similar to the positive regulator section.

The negative regulator is made up of the following components: IC2, Q2, R4, R5, R6, and R7. No zener diode is needed to set a reference voltage in this section of the circuit. The positive, regulated output voltage is used as the reference voltage source. Note that transistor Q2 is a pnp unit to account for the change of polarity.

Resistor R7 references the non-inverting input of the second op amp (IC2) to ground. Resistors R4 and R6 have identical values. This means that the voltage at their junction is 0 if the regulated V+ signal is equal to the regulated V− output voltage. This junction voltage is fed into the op-amp's inverting input through R5.

If the two output voltages are not equal, a non-zero signal is fed into the op-amp's inverting input. This causes the op-amp's output voltage to shift. This voltage controls the base current of transistor Q2. Because the V+ voltage is regulated by the other half of the circuit, changes in Q2's base current cause it to correct the V− output voltage.

One major advantage of this circuit is that even if one of the output voltages changes momentarily, the outputs automatically maintain symmetry, because the V− regulated voltage is referenced to

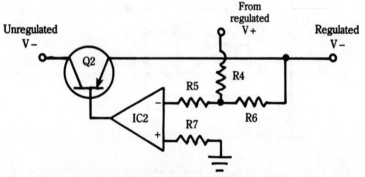

Fig. 7-20. This is the negative portion of the circuit of Fig. 7-18.

the V+ regulated voltage. This is a major advantage in many circuits, such as precision op-amp circuits.

Almost any op amps may be used in this circuit. High-grade devices aren't necessary here.

PROJECT 47: ROULETTE WHEEL

This project is a lot of fun. Figure 7-21 shows the schematic. The parts list appears in Table 7-10.

The ten LEDs should be mounted in a circle. Only one LED lights at a time. When switch S1 is briefly closed, the counter cycles through at a high rate. The lit LED appears to move rapidly around the circle. Clicks are heard through the speaker each time the LED advances position.

When the switch is released (opened), the counter gradually slows down. The effect is like a roulette wheel slowing down in rotation. Eventually the counter stops, with one LED remaining lit.

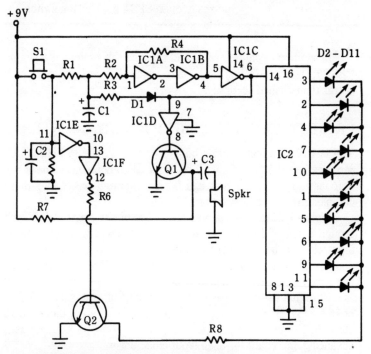

Fig. 7-21. Project 47 simulates a roulette wheel.

Table 7-10. Parts List for Project 47.

Component	Part
IC1	CD4069 hex inverter
IC2	CD4017 decade counter
Q1, Q2	npn transistor (2N2222, or similar)
D1	diode (1N914, or similar)
C1	1-μF, 25-volt electrolytic capacitor
C2	220-μF, 25-volt electrolytic capacitor
C3	4.7-μF, 25-volt electrolytic capacitor
R1	100k, ¼-watt resistor
R2	470k, ¼-watt resistor
R3, R6, R7	10k, ¼-watt resistor
R4	3.3-megohm, ¼-watt resistor
R5	1-megohm, ¼-watt resistor
R8	100-ohm, ¼-watt resistor
S1	normally open SPST pushbutton switch

In addition to games, you could also use this project as a simple "ESP tester." See if you can predict which LED will remain lit with better than chance accuracy.

PROJECT 48: SIXTEEN-STEP BINARY COUNTER

The last two projects in this book are quite simple to build. Each requires just the two ICs with no external components. These are demonstration circuits that can be incorporated easily into more complex projects and systems.

Figure 7-22 shows a simple four-stage binary counter. It is built around a pair of dual D-type flip-flop ICs (CD4013). Each time an input clock pulse is received, the count advances one step. The counter cycles through 16 steps, then repeats. Table 7-11 compares the binary count with decimal values.

Note that the count resets to zero (0000) after it passes a count of 15 (1111). Because zero is included as a count step, there are 16 steps in the count sequence. This pattern repeats continuously, advancing one count for each input clock pulse.

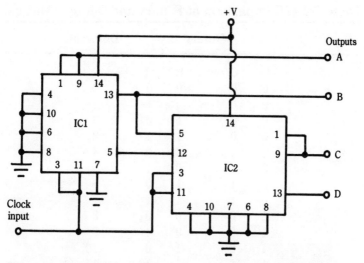

Fig. 7-22. This circuit for Project 48 is a four-stage binary counter.

The four outputs can be used to drive any digitally controllable circuitry. The count outputs also can be used as trigger signals for analog circuits.

If you would like to view the operation of this circuit, just drive an LED from each output. Be sure to use a current-dropping resistor to keep the LED from burning itself out. The simple LED modification is illustrated in Fig. 7-23. This modification should be repeated four times, once for each of the counter outputs.

Output A is the least-significant bit (LSB), and output D is the most-significant bit (MSB).

PROJECT 49: DIVIDE-BY-FIVE CIRCUIT

It is easy to divide a signal frequency by two—just pass the signal through a flip-flop. The signal to be frequency-divided is used as the flip-flop's clock signal. On each clock pulse, the flip-flop's output reverses states. Thus, it takes two complete input cycles to make up one complete output cycle. The output frequency is one-half of the input frequency.

There is no problem dividing a signal frequency by any power of two. Just cascade a number of flip-flop stages. To divide by four, use two stages. To divide by eight, use three stages. To divide by sixteen, use four stages, and so forth.

115

Table 7-11. Comparison of Binary and Decimal Values.

	Binary			Decimal
D	C	B	A	
0	0	0	0	0
0	0	0	1	1
0	0	1	0	2
0	0	1	1	3
0	1	0	0	4
0	1	0	1	5
0	1	1	0	6
0	1	1	1	7
1	0	0	0	8
1	0	0	1	9
1	0	1	0	10
1	0	1	1	11
1	1	0	0	12
1	1	0	1	13
1	1	1	0	14
1	1	1	1	15
0	0	0	0	0
0	0	0	1	1
0	0	1	0	2
0	0	1	1	3

However, what if you need to divide the signal frequency by an intermediate value? In this project you will be dividing by five.

The solution is to use JK-type flip-flops. The circuit can be forcibly reset at any point.

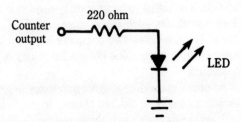

Fig. 7-23. LEDs can be added easily to the outputs of the circuit in Fig. 7-22.

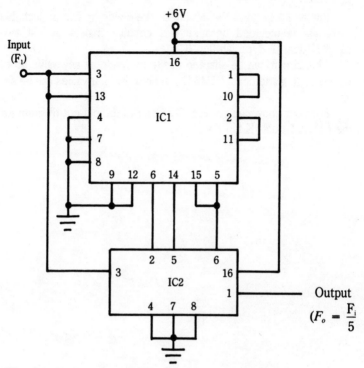

Fig. 7-24. Project 49 divides the input frequency by five.

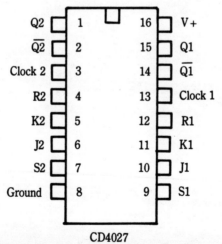

CD4027

Fig. 7-25. The CD4027 dual JK-type flip-flop IC is used in the circuit of Fig. 7-24.

Figure 7-24 shows the circuit for the divide-by-five project. No parts list is included because this circuit consists of just two CD4027 dual JK-type flip-flop ICs.

This circuit can be adapted easily to divide by any value up to 16. By adding additional CD4027s, higher division values can be set up.

For your convenience, Fig. 7-25 shows the pin-out diagram for the CD4027 dual-JK flip-flop.

Index